D1618808

Geophysics and Meteorology at the University of Leipzig

On the Occasion of the 100th Anniversary
of the Foundation of the Geophysical Institute
in 1913

Werner Ehrmann and Manfred Wendisch (eds.)

Geophysics and Meteorology at the University of Leipzig

On the Occasion of the 100th Anniversary of the Foundation of the Geophysical Institute in 1913

with contributions by
Michael Börngen, Peter Hupfer, Christoph Jacobi,
Franz Jacobs, Michael Korn, Sigurd Schienbein,
Hans-Jürgen Schönfeldt, Dietrich Sonntag,
Ludwig A. Weickmann, and Manfred Wendisch

LEIPZIGER UNIVERSITÄTSVERLAG GMBH 2013

Bibliografische Information der Deutschen Nationalbibliothek
Die Deutsche Nationalbibliothek verzeichnet diese Publikation in der deutschen Nationalbibliografie; detaillierte bibliografische Daten sind im Internet über http://dnb.d-nb.de abrufbar.

Das Werk einschließlich aller seiner Teile ist urheberrechtlich geschützt. Jede Verwertung außerhalb der engen Grenzen des Urheberrechtsgesetzes ist ohne Zustimmung des Verlages unzulässig und strafbar. Das gilt insbesondere für Vervielfältigungen, Übersetzungen, Mikroverfilmungen und die Einspeicherung und Verarbeitung in elektronischen Systemen.

Special issue 2 of "Leipziger Geowissenschaften"
Special issue 2 of "Wissenschaftliche Mitteilungen aus dem Institut für Meteorologie der Universität Leipzig"

With support of
Deutsche Geophysikalische Gesellschaft e. V., Arbeitskreis "Geschichte der Geophysik"
Deutsche Meteorologische Gesellschaft e V., Fachausschuss "Geschichte der Meteorologie"

Title Figures:
Collm Geophysical Observatory (Postcard, E. Assmus, Leipzig, around 1938); Norwegian stamp on the occasion of the 100th anniversary of the birth of Vilhelm Bjerknes published in 1962; Volcano Merapi. Finite-element model and electrical resistivity tomography; Intensification of climate change due to the effect of anthropogenic aerosol particles on clouds ("indirect aerosol effect", in watts per square metre) given as temporal mean of the year 2003.

© Leipziger Universitätsverlag GmbH 2013
Redaction: Michael Börngen and Franz Jacobs
Layout/Production: UFER VERLAGSHERSTELLUNG, Leipzig

ISBN 978-3-86583-742-4

Content

Preface

The early years of the 20th century saw a rapid and vigorous development of the natural sciences and technology in Germany, and the University of Leipzig was involved in the fast moving evolution. In this quickly changing environment, the Geophysical Institute was established at the University of Leipzig on January 1, 1913. At that time geophysics and meteorology were considered subsections of physics and geosciences. They were divided into Physics of the Solid Earth (Geophysics), Physics of the Atmosphere (Meteorology), and Physics of the Hydrosphere (Oceanography). The goal of the new Geophysical Institute was the joint treatment of all three scientific branches, beginning with meteorology. Although traditions in geosciences and meteorology in Leipzig reach much further back in time, the foundation of the new Geophysical Institute promoted the rapid advancement of these disciplines in Leipzig itself and beyond. The institute was the first in Germany to deal specifically with meteorology and the second, after Göttingen, devoted specifically to geophysics.

Over its history, the institute has experienced several periods of heartening growth followed by phases of frustrating decline. Apart from the war years, we must particularly mention in this context the various restructurings and the closure of study programmes by the governments. On the other hand, teaching and research in meteorology and geophysics was strongly invigorated after the German unification by the foundation of the Institute for Meteorology and the Institute of Geophysics and Geology.

Since the foundation of the Geophysical Institute one hundred years ago, meteorology and geophysics have evolved to become highly specialised and autonomous branches within modern natural science, and therefore have been organized into two separate institutes (Institute for Meteorology and Institute of Geophysics and Geology) within the Faculty of Physics and Earth Sciences since 1993. These have developed into nationally and internationally respected institutes, and have become core research and teaching units within the faculty and the university as a whole. A large number of students study meteorology and geophysics, receiving their bachelor/master/diploma degrees or doctorates from the University of Leipzig. The two institutes also offer courses for students of other disciplines.

Both institutes are characterised by high-level national and international research activities, which are documented in a large number of peer-reviewed publications. There are also numerous collaborations with non-university research institutions such as the Leibniz-Institut für Troposphärenforschung, TROPOS (Leibniz Institute for Tropospheric Research), the Helmholtz-Zentrum für Umweltforschung, UFZ (Helmholtz Centre for Environmental Research), and the Helmholtz-Zentrum Potsdam Deutsches GeoForschungZentrum, GFZ (Helmholtz Centre Potsdam, German Research Centre for Geosciences). Furthermore, many fruitful co-operations exist with other national and international universities and with scientific institutes and agencies, for example Deutscher Wetterdienst, DWD (German Weather Service).

This volume outlines the important stages in the history of geophysics and meteorology at the University of Leipzig from the Geophysical Institute in 1913 to the Institute of Geophysics and Geology and the Institute for Meteorology in 2013.

Manfred Wendisch
Institute for Meteorology

Werner Ehrmann
Institute of Geophysics and Geology

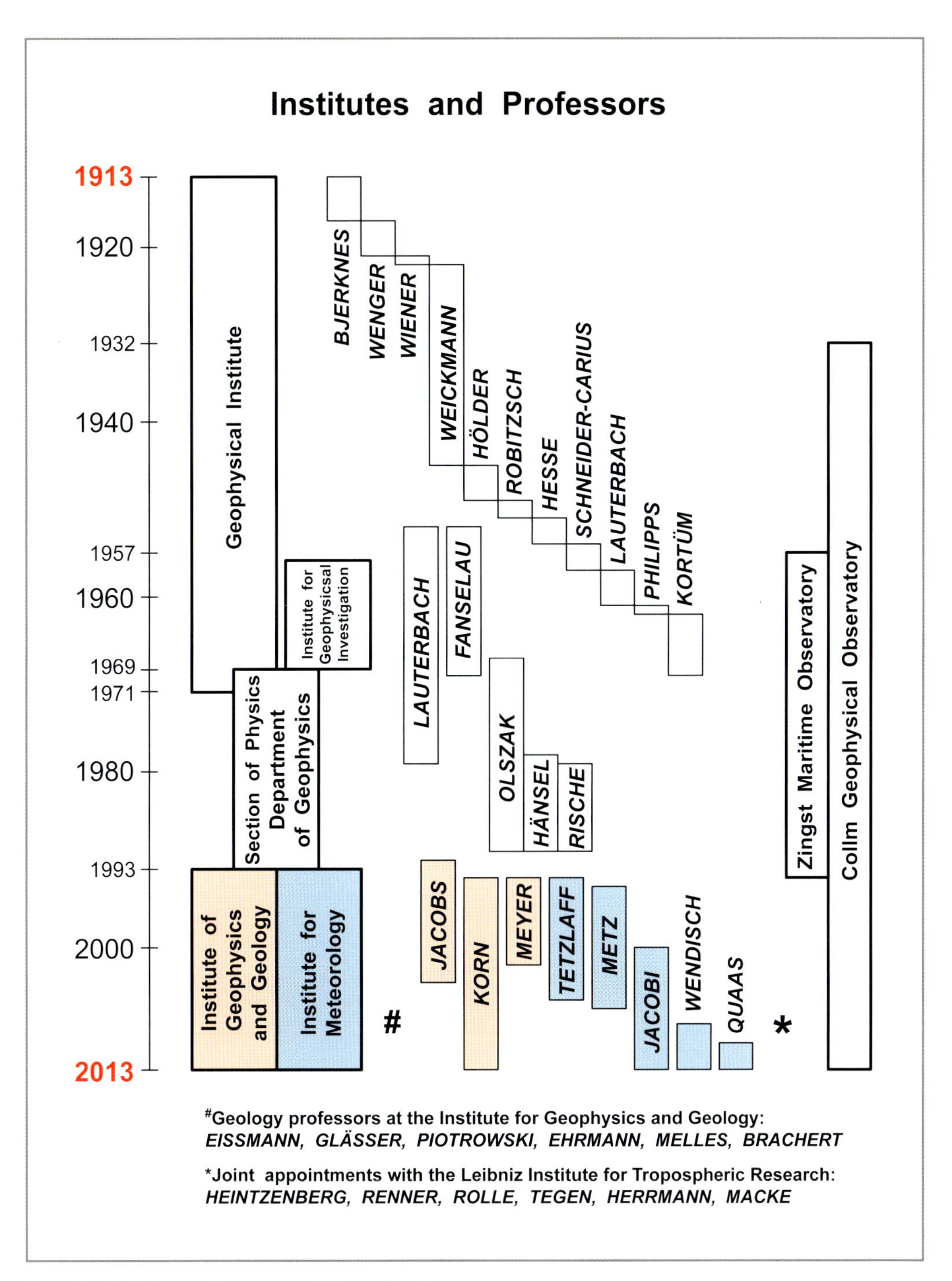

Compiled by Werner Ehrmann and Franz Jacobs.

The Geophysical Institute 1913–1923: The early Years

Michael Börngen and Ludwig A. Weickmann

Traditions

The first steps were taken towards establishing a Geophysical Institute at the University of Leipzig in 1910. By then, however, geophysics (the term itself originated at the end of the 19th century) and meteorology could already look back on a remarkable tradition.

In the first half of 19th century, the study of geomagnetism – then linked closely with meteorology – was of special interest. Carl Friedrich Gauss and Wilhelm Eduard Weber established a geomagnetic observatory in 1833 in Göttingen which soon became the headquarters of the "Göttinger Magnetischer Verein", bringing together observatories from around the world to make observations on specific days. The University of Leipzig observatory was a member from 1834 to 1841, when it was headed by August Ferdinand Möbius, discoverer of the Möbius strip. When Weber, after losing his job along with six other professors at Göttingen for political reasons, took over the chair of physics at the University of Leipzig from 1843 to 1849, he founded a magnetic observatory near Augustusplatz. The declination and inclination of the Earth's magnetic field were measured here for several years (Fig. 1.1) before the observatory moved to Talstraße 35.

104 D'ARREST, BESTIMMUNG DER DECLINATION IM

1850. Westliche Declinationen.

Datum.	Mittlere Zeit.	Declination.	Mittlere Zeit.	Declination.
Oct. 2	2h 25′	15° 42′ 1″	21h 10′	15° 36′ 43″
3	3 5	15 40 17	21 30	15 31 37
4	2 35	15 40 41	21 0	15 33 7
5	2 15	15 47 42	....	......
6	....	......	21 10	15 43 17
7	3 0	15 45 7	21 25	15 40 30
8	3 3	15 43 59	21 17	15 37 26
9	3 0	15 45 20	....	......
10	3 0	15 44 56	21 30	15 40 28
11	2 51	15 44 22	21 26	15 35 5
12	....	......	21 29	15 34 37
13	2 10	15 43 13	21 17	15 35 19
14	3 10	15 41 6	21 4	15 34 12
15	2 40	15 46 6	....	......
16	....	......	21 20	15 38 23
17	2 47	15 40 45	21 25	15 35 24
18	3 5	15 42 30	21 30	15 36 23
19	....	......	21 28	15 37 26
21	3 5	15 38 21	21 15	15 36 2
22	2 15	15 42 26	21 20	15 35 57
23	3 15	15 40 24	21 19	15 38 20
24	....	......	21 45	15 39 18

Figure 1.1: Declination values in Leipzig in October 1850 (From d'Arrest 1851).

The geologist and palaeontologist Hermann Credner (Fig. 1.2), director of the "Geologische Landesuntersuchung von Sachsen," surveying Saxony's geological make-up, and later director of the Geological-Palaeontological Institute at the University of Leipzig, worked to advance earthquake research for many decades. In the 1870s, attention turned towards perceptible earthquakes in Saxony, and Credner began to collect observations and evaluate them. To assist him in the task, he created the *Saxon earthquake commission* in 1898, consisting of honorary members, and in 1899, together with Georg Gerland, a permanent commission for international earthquake research, whose task it was to set up an international seismologic society. This society was finally established in Strasbourg in 1903, eventually developing into today's IASPEI (International Association of Seismology and Physics of the Earth's Interior). In order to supplement his phenomenological observation, Credner set up an earthquake observatory in the basement of the institute in Talstraße 35 in 1902. The step was made in close co-operation with Emil Wiechert, who in 1898 had become the holder of the world's first chair for geophysics, in Göttingen. The earthquake observatory, supervised by Franz Etzold, was equipped with a pendulum seismometer designed by Wiechert to Credner's specifications with a mass of 1.1 tonnes. The approximately 250-fold magnification of the *most sensitive apparatus in the world* (Wiechert) allowed earthquakes in the Saxon-Vogtland area to be recorded in sufficient detail and more distant earthquakes to be registered (Fig. 1.3). The site of the earthquake station is marked today by a commemorative tablet, in the courtyard of the Institute of Geophysics and Geology.

Figure 1.2: Hermann Credner (1841–1913).

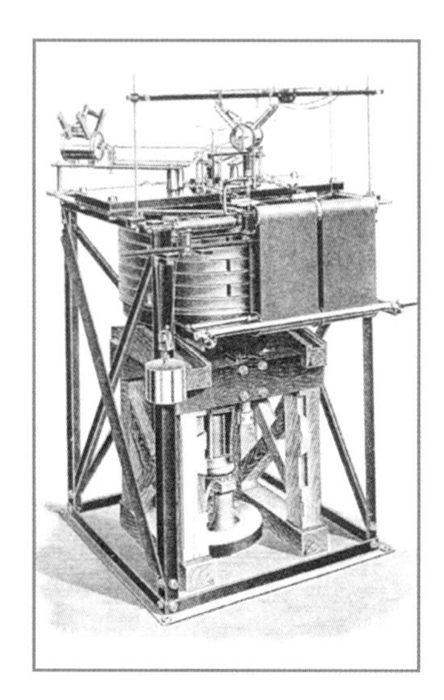

Figure 1.3: Wiechert seismograph.

The traditions of meteorology in Leipzig can be traced even as far as 1507, the year that the "Decalogium [...] de Metheorologicis [...]" was published (Fig. 1.4). This work is commonly held to be the first meteorology textbook. The author was Virgil Wellendarffer (sometimes spelled Wellendorffer or Wallendorffer), *Magister* in philosophy and *Baccalaureus* in theology and Rector of the University of Leipzig in 1502. Following the invention of a number of meteorological instruments in the 17th century, the weather had begun to be "measured" in Leipzig by 1728. When Heinrich Wilhelm Brandes took over the chair of physics at the university in 1826, he presented a dissertation which contained the first, if still very simple, synoptic weather maps. This study,

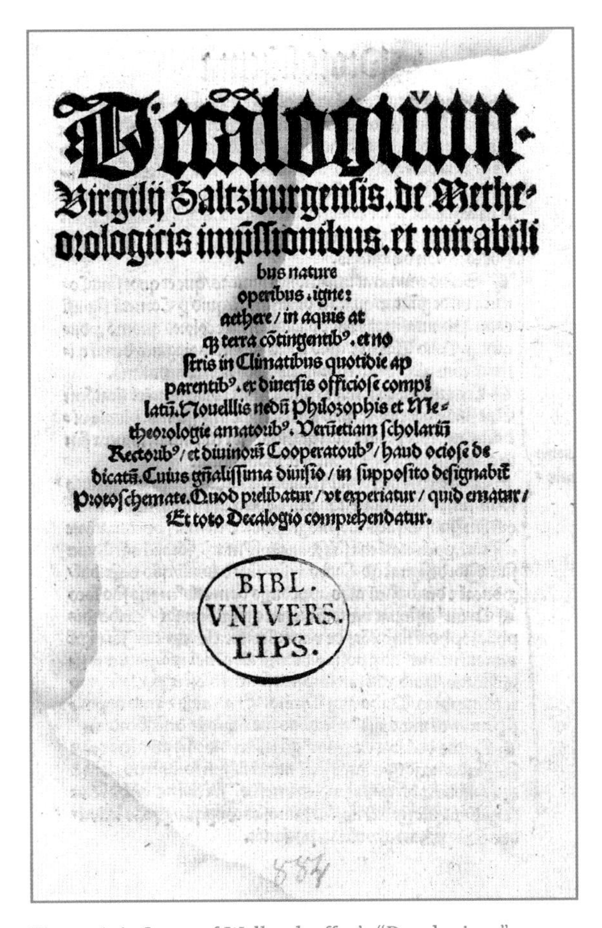

Decalogium.
Virgilij Saltzburgensis. de Metheorologicis impressionibus. et mirabilibus nature
operibus. igne:
aethere / in aquis at
que terra cotingentib⁹. et no
stris in Climatibus quotidie ap
parentib⁹. et diversis officiose compi
latu. Novellis nedum Philozophis et Me
theorologie amatorib⁹. Verumetiam scholarium
Rectorib⁹ / et divinorum Cooperatorib⁹ / haud ociose de
dicatum. Cuius generalissima divisio / in supposito designabitur
Protoschemate. Quod prelibatur / ut experiatur / quid enatur /
Et toto Decalogio comprehendatur.

BIBL.
VNIVERS.
LIPS.

Figure 1.4: Cover of Wellendarffer's "Decalogium".

combined with Brandes' earlier *contributions to weather theory* written in Breslau but published in Leipzig in 1820, was the beginning of synoptic meteorology, a new branch within the discipline. In the last years of his life, Brandes did significant work laying the foundations for the establishment of the Physical Institute at the University of Leipzig in 1835.

In the 1870s, the astronomer and director of the Leipzig astronomical observatory Carl Christian Bruhns contributed significantly to worldwide cooperation among meteorologists by being the main initiator of an international conference of meteorologists held in Leipzig in 1872 and the foundation stone for the International Meteorological Organization, the forerunner of today's World Meteorological Organization (WMO). Bruhns was also among the founders and the first heads of a meteorological measurement network in Saxony (beginning December, 1863). He even managed to establish a *meteorological office of weather prediction* (1878) in the very centre of Leipzig. His important contribution to triangulation measurements in Saxony within the framework of the Central European degree measurement should also be mentioned.

Bruhns' successor, Heinrich Bruns, contributed significantly to the advancement of geosciences with his seminal work about the figure of the Earth (Bruns 1878).

The Bjerknes Era

At the end of 19th century, meteorology was regarded more or less as a realm of observation and statistics, and a playground for instrument development. Some scientists realized that the emphasis on climatology as the dominating subject did not provide a response to the growing public demand for weather forecasting nor meet the requirements of the early aviation industry, which at that time mostly involved balloons and airships. Other scientists mocked meteorology as being more or less on a par with astrology, saying that it should aim to become "meteoronomy", an exact science like astronomy. Although the name meteorology remained, there nevertheless developed a trend to introduce more physics and mathematics. Among the first important scientific steps was the application of thermodynamic laws to atmospheric phenomena. While this helped much in understanding local or short range developments, it did not lead very far in understanding dynamic processes and such in solving forecasting problems.

Figure 1.5: Otto Wiener (1862–1927).

Otto Wiener (Fig. 1.5), director of the Physical Institute at the University of Leipzig, was among the first scientists who realized the necessity of dynamics for further progress and saw that this called for a specialised institution at a university. Wiener had followed the work of the Norwegian physicist Vilhelm Bjerknes (Fig. 1.6) and considered him as one of the most competent scientists in the field of dynamics at the time. As early as 1910, Wiener had initiated the establishment of such an institute in Leipzig. It was a lucky coincidence that funds could be made available to employ staff and pay for office space and that Bjerknes accepted the offer to take over the headship of the new institution.

The new institute was named "Geophysikalisches Institut" (Geophysical Institute) in anticipation of the solid earth and the hydrosphere also being included in the teaching and research programme later on.

Figure 1.6: Vilhelm Bjerknes (1862–1951).

This institute, the second oldest geophysics institute in Germany after Göttingen, officially opened on 1 January 1913. *During the foundation of the institute, I steered towards the solution of only one problem, the problem of the weather forecast*, said the first director Bjerknes later. He wanted to solve the problem with the help of theoretical thermodynamics and hydrodynamics instead of the statistical methods that had been used before. The basis for this kind of approach to weather forecasting, which is still valid today, was an innovative publication by Bjerknes from 1904.

Bjerknes inaugural address on 8 January 1913 contained his famous statement that *I would be more than glad if I could take the work so far that I could predict only tomorrow's weather after even year-long calculations. If just the calculations were correct, then the scientific victory would have been achieved. Meteorology would then become an accurate science, a real physics of the atmosphere. And if we were so far only, already practical consequences would arise.*

The newly established institute was housed in the garden annexe at Nürnberger Straße 57 (Fig. 1.7) and initially consisted of just 6 small rooms on the first floor. By the following year, however, the whole floor was available and was eagerly taken over.

The research staff at the institute included 2 assistants and 2 or 3 auxiliary assistants. Also among the scientific staff were Theodor Hesselberg and Harald Ulrik Sverdrup, and, later, Jacob Bjerknes and Halvor Solberg. These scientists were private assistants of Bjerknes, paid for by an American Carnegie Institution endowment. In addition, another secretary and a draughtsman (Bjerknes placed great value on this) were employed at the institute.

Bjerknes was supported by his assistant Robert Wenger, peviously head of the observatory on Tenerife, who had already been involved in the comprehensive preparatory work and had to this end spent several months in Norway. Apart from his research, Wenger's duties comprised the management of the institute and the organization of teaching in practice. Wenger completed his post-doctoral habilitation qualification in the summer of 1917 with a thesis on methodological errors in aerologic observations. He gave his trial lecture on the current state of foehn theory on July 17 of that year.

In the summer of 1914, the Russian physicist Aleksandr A. Friedmann (also spelled Fridman), known as a cosmologist, came to the Geophysical Institute to learn about Bjerknes' working methods. Together with Hesselberg, he published a scientific treatment of the scale of the meteorological elements and their spatial and temporal derivatives (Hesselberg, Friedmann 1914).

Figure 1.7: Nürnberger Straße 57, domicile of the Geophysical Institute from 1913 to 1917 (Photo by 2012).

Bjerknes himself became a member of the Saxon Academy of Sciences, publishing several works in its *treatises* and was also awarded the honorary title of *Geheimrat*, or Privy Councillor.

Bjerknes and his successors as director of the Geophysical Institute published "Veröffentlichungen des Geophysikalischen Instituts der Universität Leipzig", a journal consisting in two series. Series I (on synoptic representations of atmospheric conditions) contained analyses of international observations from aerological ascents using Bjerknes' methods. Two volumes of this series with a total of 10 issues were published, most of them edited by Bjerknes, but with the last two edited by Wenger (Fig. 1.8). The last issue appeared in 1918. The deteriorating financial situation made further printing of such costly map series impossible. Series II ("Spezialarbeiten aus dem Geophysikalischen Institut und Observatorium") was published from 1913 up to 1970 (most recently in the Akademie-Verlag, Berlin), and contained above all dissertations written at the Institute.

In 1916, Luise Lammert (Fig. 2.5) joined the Geophysical Institute while still completing her academic studies. Beside Marie Dietsch, who worked at the institute for a short time – Lammert is one of Germany's first university-trained female meteorologists, remaining at the institute until 1935. In 1919, she was awarded a doctorate for her work on the average state of the atmosphere during a southerly foehn (Lammert 1920).

Bjerknes spent almost five very successful years in Leipzig, filled with lecturing (Fig. 1.9), organizing the institute and thus laying the foundation for a modern view on meteorology. He was able to continue his work despite the outbreak of World War I, even though living conditions became increasingly difficult.

Veröffentlichungen
des Geophysikalischen Instituts der Universität Leipzig
Herausgegeben von dessen Direktor
V. Bjerknes

Erste Serie

Synoptische Darstellung atmosphärischer Zustände

Jahrgang 1911

Heft 1

Zustand
der Atmosphäre über Europa
am 1., 2., und 3. März 1911

bearbeitet von

R. Wenger

B: Erläuterungen und Tabellen

Geophysikalisches Institut
der Universität
LEIPZIG
S/1465

Leipzig 1916

Figure 1.8: Cover of "Veröffentlichungen des Geophysikalischen Instituts, Erste Serie".

82. DIE WIRBELRÖHREN IN DER ATMOSPHÄRE.
Man kann sich leicht einen allgemeinen Ueberblick über die Wirbelverteilung in der Atmosphäre schaffen.
Man Wo man am Erdboden auf der nördlichen Halbkugel zyklonische Horizontalzirkulation hat, steigen Wirbelröhren aus dem Boden empor. Wo antizyklonische Horizontalzirkulation vorliegt, steigen sie zum Boden herab. Auf der südlichen Halbkugel umgekehrt. Jetzt ist die allgemeine Windverteilung: Ostwind im Äquatorialgebiet zwischen den Rossbreiten, Westwind zwischen Rossbreiten und Pol. Dies gibt zwei antizyklonische Zonen längs den Rossbreiten, zwei zyklonische um den Pol. Also: positive Wirbelröhren steigen auf aus dem nörd-

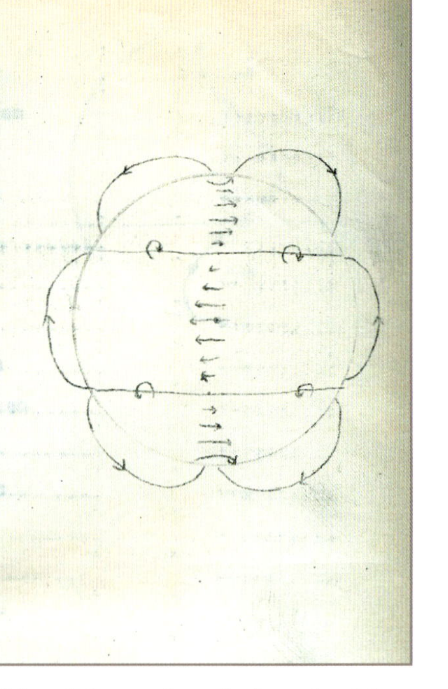

Figure 1.9: Example from Bjerknes' lecture notes of 1916/17 (*Selected chapters from the dynamics of the atmosphere and the sea*), Section 82 (*Vortex tubes in the atmosphere*). Doubles space type script in the left column, idealized sketch of the global circulation (in pencil) in the right one.

Most of Bjerknes' students were recruited for military service, and three of them – H. Petzold, H. Behrend and C. Boerner – were killed in the war. During the war years, several flying schools were set up around Leipzig and Wenger gave courses in meteorology for the pilots. In late 1916, he himself was conscripted and served as a meteorologist at Grand Headquarters. There, he wrote a review for the army weather service issued by the Commanding General of the air force, on contributions to editing aerological observations and their exploitation in the weather forecast, a treatise which became fundamental for military weather services.

Contrary to some later misinterpretation, according to which Bjerknes had been requested to support the German military command with daily weather forecasts etc., H. U. Sverdrup states: "... *I may mention that no demands for participation in meteorological research related to military problems were made on Bjerknes or on his two or three Norwegian assistants, who continued their strictly academic work and were paid by the grant from the Carnegie Institution in Washington even after the United States had entered the war*" (Sverdrup 1951, p. 220).

Bjerknes demanded an improvement of the room situation as soon as possible despite the war. Suitable rooms were found in the building previously occupied by an institute for deaf-mutes, located at Talstraße 38 (Fig. 2.3). The building had been taken over by the university and completely refurbished. Luise Lammert reported that, *half of the 3rd floor was occupied by the Geophysical Institute, and at the beginning of the summer semester 1917 the removal to Talstraße took place. Here, there were large well-lit rooms available, and the institute was also furnished with its own lecture theatre, geophysical lectures having so far been held in the Physical and Mathematical Institute. In consideration of the geophysical practical courses to be held, the lecture room is equipped with broad single tables, in order to make drawing of large weather charts possible. The flat roof of the house has also been put at the disposal of the Geophysical Institute, so that pilot balloon observations can be carried out from there, and meteorological apparatus may be set up later. One of the attic rooms is provisionally equipped as a laboratory with calibration devices and a lathe* (Lammert 1932). Other rooms in the building had been given to the Mineralogical-Petrographic Institute.

At the end of September 1917, Bjerknes, in amicable agreement with the university, resigned the directorship of the institute to take up a chair in Bergen. He was accompanied by his son Jacob and his two Norwegian assistants, Sverdrup and Hesselberg. Thus, as Weickmann commented, Bjerknes' work *experienced its culmination not in Leipzig, but in Bergen, where the polar front theory and the wave theory of the cyclones were developed in direct connection to the Leipzig work, and where the fundamentals were developed for the most important and most successful method of modern weather chart analysis and weather forecast, which is nowadays widespread throughout the world* (Weickmann 1938, p. 7–8).

In his speech on the occasion of the 25th anniversary of the Geophysical Institute in 1938, nevertheless, Bjerknes himself said, *however small the contribution of Leipzig to the final success in Bergen may appear when viewed from outside, yet this contribution is not only great, but has been indispensable for the final success* (Bjerknes 1938, p. 60).

Wenger

Before his departure, Bjerknes ensured that Robert Wenger (Fig. 1.10), his colleague of many years and a man qualified like no other to continue his work, was released from military service in September 1917. Wenger was immediately appointed both as acting director of the Geophysical Institute and extraordinary professor. His inaugural lecture on the predetermination of the weather took place on July 20, 1918, and in summer 1920 he became Director of the Geophysical Institute and was appointed full professor.

Since Wenger attached great importance to the comprehensive observational material from military weather and kite stations, he agreed with the directors of

Figure 1.10: Robert Wenger (1886–1922).

the army weather service that meteorological and aerological material would be sent to the institute continuously. The material contained data from kite ascents and pilot balloon observations from October 1917 until October 1918, and Luise Lammert and Marie Dietsch used some of these observations for their joint paper on normal wind and friction force at 1000 m height (Lammert; Dietsch 1921).

Soon after the end of war, Wenger was invited by Bjerknes to Norway from July to October 1919 so that he could see the progress made in Bergen in the intervening time. In May 1920, Wenger turned down an invitation to go to Kristiania as director of the Norwegian weather service.

In line with Bjerknes' principle "interpretation not observation," the equipment at the institute in the early years consisted of only a few meteorological instruments. Wenger was anxious to maintain the traditions of Bjerknes at the Geophysical Institute, but also wanted to sustain practical meteorology. He was of the opinion that good equipment and a continuous observational program were essential both for research and for the practical training of students. He managed to procure meteorological instruments for the institute from army supplies, using funds granted especially for that purpose. This meant that he could incorporate instrumental and, as far as was possible in post-war conditions, aerological work into the curriculum more intensively. One result was that active relations developed between the institute and the aeronautic observatory in Lindenberg.

Wenger had a strong interest in radiation measurement, and so he began preliminary testing in preparation for the construction of a new type of actinometer according to the Wien-Lummer black body model, and purchased a Michelson actinometer and a Jensen radiation apparatus.

In October 1919, Wenger set up a daily weather service at the institute, since he expected great advantages from continuous monitoring of the weather. The weather service was intended to be part of the training for young meteorologists, but it was also integrated into scientific research, as interesting events noticeable on the daily weather chart, were to be examined in detail. A daily discussion of the weather conditions, with analysis and discussion of the weather map, became compulsory for all institute members from the summer semester of 1920 on.

In October 1920, Wenger took over the earthquake station, which up to then had been part of the Geological-Palaeontological Institute at Talstraße 35. The Geophysical Institute also took over responsibility for the seismological station in Plauen run by Ernst Weise. Both stations housed Wiechert horizontal seismometers, but at the time of the transfer, the Leipzig seismograph was in urgent need of repair. The necessary funding was not forthcoming, however, with the result that the apparatus had to be provisionally taken out of service in 1921.

Despite the loss of the seismograph, Franz Kossmat, Director of the Geological-Palaeontological Institute, nonetheless paid sufficient attention to geophysics. Of special importance is the gravity map of Central Europe (Fig. 1.11) he compiled in 1920 together with Hans Lissner. The map gave insight into the density distribution of the earth's crust and made statements possible regarding the formation and genesis of the examined area. Kossmat produced numerous publications on the geological-geophysical problems of gravimetry, isostasy and orogeny.

There can be no doubt that the co-operation between Leipzig and Bergen initiated under Wenger would have produced very fruitful results, but developments in Leipzig were again suddenly interrupted when Wenger died suddenly of complications of influenza on January 20, 1922 at the age of just 36. Vilhelm Bjerknes and Hugo Hergesell both paid detailed tribute to his achievements, and Otto Wiener, always closely associated with the Geophysical Institute, provisionally took over its direction.

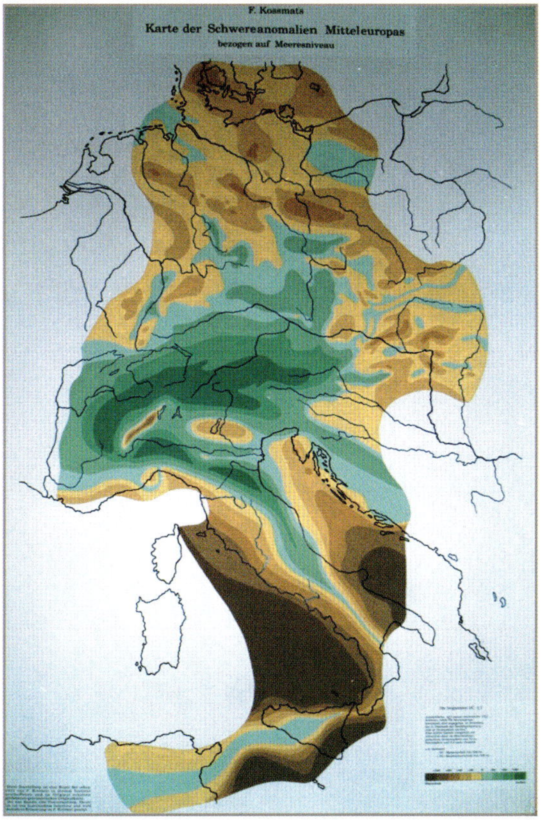

Figure 1.11: Map of gravity anomalies in Middle Europe after Kossmat, reconstructed by Horst Rast.

Wenger had already planned a flight weather service. Shortly after his death, there came the opportunity to use the institute weather service for air traffic purposes. As requested by the aviation authority and the air police, the institute declared itself ready to give out weather forecasts by telephone and to make daily pilot balloon ascents to study the lower air layers.

It is assumed that also Wenger – along with Kossmat – was involved in the foundation of the German Seismological Society proposed by Emil Wiechert, even though he died before it was officially established on September 19, 1922. The act of establishment fittingly took place at the Geophysical Institute of the University of Leipzig during the centenary anniversary conference of the

Nr. 1 Montag, 18. September 1922

TAGEBLATT

der

HUNDERTJAHRFEIER
DEUTSCHER NATURFORSCHER
UND AERZTE IN LEIPZIG

vom 17. bis 24. September 1922.

Abteilung 10a:
Geophysik.
(einschließlich Seismologie).
Einführende: Prof. Dr. O. Wiener, Linnéstr. 4 (23 120); Prof. Dr. Bauschinger, Stephanstr. 8 (27 094); Schriftführer: Dr. Lammert, L.-Gohlis, Fritzschestr. 29.
Sitzungsort: Hörsaal für Geophysik, Talstraße 38.
Treffpunkt: Deutsches Haus, Königsplatz.
Vorträge für die Abteilung Geophysik:
Dienstag, den 19. September, nachm. 3 Uhr:
Sitzung der „Versammlung der Seismologen" im Hörsaal für Geophysik

Figure 1.12: Documents of the founding of the German Seismological Society.

GDNÄ, the Society of German Natural Scientists and Physicians (Fig. 1.12). In place of Wenger, Otto Wiener as the provisional Director of the Geophysical Institute chaired the foundation meeting with Luise Lammert functioning as secretary. The meeting agreed that the *promotion of theoretical and practical seismology and related questions* were the Society's object, but only two years later the programme was expanded and the society was renamed Deutsche Geophysikalische Gesellschaft (DGG), the German Geophysical Society.

2 The Geophysical Institute 1923–1945: Heyday and Decline

Ludwig A. Weickmann and Michael Börngen

On October 1, 1923, Ludwig F. Weickmann (Figures 2.1, 2.2) from Munich took over the direction of the Geophysical Institute at the same time as he took up his appointment to the Chair for Geophysics. Thanks to his unusual talent in organizing science and his outstanding personality as a university teacher, the institute was to develop into a teaching and research institution of international standing.

Weickmann wrote later, that he *was determined to continue Bjerknes' tradition in Leipzig, to make use of the progress that had achieved in Norway since 1917 with the "Bergen Methods" for the benefit of the slowly developing German air traffic and the meteorological advisory service. At the same time, I wanted to push the institute forward in the sense of incorporating other geophysical branches into teaching and research. Wenger had already made a successful start with that. He had built an actinometer in a small institute-owned workshop [...], had taken over the Wiechert seismograph from the Geological Institute [...], and had arranged for a good meteorological station of I. order. It was my goal to proceed in this direction. However, financial difficulties hampered these efforts* (Weickmann 1938, p. 9).

Figure 2.1: Ludwig F. Weickmann (1882–1961).

The Geophysical Institute (Fig. 2.3) had actually been an institute for theoretical meteorology under Bjerknes. Wenger introduced practical meteorology and Weickmann then expanded his research and teaching into all fields of geophysics. He strove to acquire meteorological, aerologic and geophysical instruments, and so retaining possession of the Wiechert seismograph was one of the first of Weickmann's tasks. It was secured after supporters and friends of the university had donated a considerable amount of money. The apparatus was thoroughly overhauled in Göttingen and reinstalled in January 1925 in its old home with the assistance of Gerhard Krumbach from the national institute for earthquake research in Jena. However, microseismic vibrations caused by traffic and machines had increased in the meantime to such a degree that fine deflections in the seismograms could

Figure 2.2: Walther Nernst (left), Nobel Prize laureate in chemistry, Werner Heisenberg (from behind), Nobel Prize laureate in physics, and the chemist Karl Friedrich Bonhoeffer (right) at a meeting with Ludwig Weickmann in his apartment.

Figure 2.3: Talstraße 38, domicile of the Geophysical Institute from 1917 to 1943. Meteorological measuring instruments near the top floor windows at the corner of the building (cf. Figure 2.17).

no longer be distinguished. A transfer of the seismograph to an observatory outside the city's environs became absolutely essential.

Weickmann also held lectures on seismology, geomagnetism and other topics, and he extended the geophysical section of the library by taking over the earthquake-related stock from the Geological-Palaeontological Institute's library. *Only after this did the institute became a true geophysical institute, whereas before in reality it had been a meteorological institute* (Lammert 1932).

Meteorology nevertheless remained Weickmann's domain. Under his direction, the institute became the home of the Leipzig School of Meteorology for nearly a quarter century. The cordial contact with Bjerknes and his Scandinavian colleagues resulted in a fruitful cooperation in the following years, with reciprocal visits. For example, Tor Bergeron worked in Leipzig in 1923/24 following an invitation from Weickmann. Bergeron, in cooperation with the Czech Gustav Swoboda (later a member of the International Meteorological Organization Secretariat and the first Secretary General of World Meteorological Organization), was engaged in an extended study of polar front phenomena. One of the results was that Bergeron, partly in co-operation with the Bergen staff, developed a classification of the newly discovered frontal phenomena. In a postcard to Jacob Bjerknes, he entitled it "Signaturae Lipsienses". The symbols for cold front, warm front and occlusion (subdivided into upper level and lower level fronts) were first published in *Waves and Whirls on a quasistationary boundary area over Europe. Analysis of the weather period 9–14 October 1923* (Bergeron, Swoboda, 1924). The symbols are still in use today (Fig. 2.4).

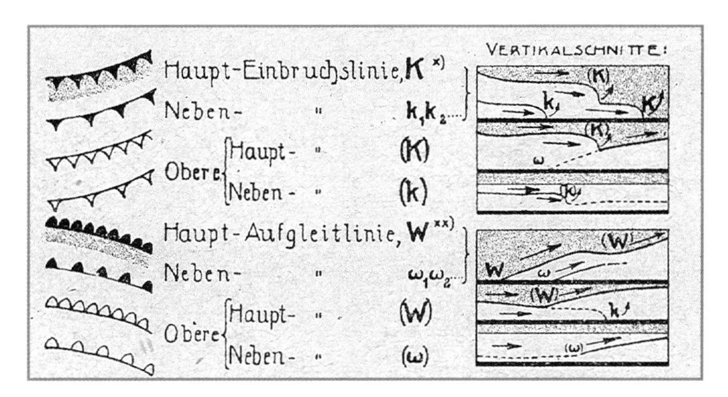

Figure 2.4: Card signatures/signs for representation of the results of front and air mass analyses ("Signaturae Lipsienses").

Air pressure waves and the theory of symmetry points were a specific research area at the Geophysical Institute. Weickmann's inaugural lecture on November 3, 1923 was on *Points of symmetry in air pressure*. After he had become a member of the Saxon Academy of Sciences in 1924, Weickmann's first contribution in October 1924 was his treatise *Waves in the Air*, for which Luise Lammert (Fig. 2.5) had also written a chapter. Numerous other studies have been carried out in the institute on these issues, building on this and other foundation-laying work by Weickmann. As a result, air pressure was found to have rhythms over several weeks, and their superposition sometimes exhibited almost mirror-

Figure 2.5:
Ludwig Weickmann and
Luise Lammert.

image behavior around a "point of symmetry", even over several months. The hope of basing forecasting on such features could, unfortunately, not be realized in practice.

Despite this limitation, the systematic investigation of mirror points as a consequence of quasi-harmonic oscillations of the atmosphere has provided a number of important results, and has led to a wealth of impulses, which enhanced common meteorological knowledge and inspired further research (Philipps 1962, p. 122).

The members of staff at the institute published their scientific results, often in form of dissertations, in the "Veröffentlichungen des Geophysikalischen Instituts der Universität Leipzig, Zweite Serie", which was established by Bjerknes. The institute's earthquake reports were printed in the Saxon Academy of Science's own publication series.

For Weickmann, the importance of aviation and the need to promote it were beyond question. He and other members of the Geophysical Institute repeatedly took part in scientific free balloon flights, during which they took scientific measurements, mainly of temperature, moisture and radiation. One of Weickmann's students, the pilot Fritz Holm Bielich, had to use an airplane to investigate the influence of city turbidity on visibility and solar radiation for his dissertation.

Soon after Weickmann had started his activities in Leipzig, he managed to get a radio communication system for the institute that allowed weather observations from European, Atlantic and American regions to be received. Two radio operators (Ernst Moritz Arndt and Clemens Auerbach) were engaged to record the data and to draw 3 or 4 weather charts per day. The radio operators received the data transmitted by radio from inside Germany and from abroad, and recorded it directly onto the weather charts. The detailed analysis and assessment of these maps during the daily weather discussions was an excellent exercise for the members of staff and students at the institute. It also was a perfect link between theory and practice: as a result of the practicing the weather forecasting, new ideas arose for specific investigation in the area of synoptic meteorology and aerology, and these were treated in subsequent research at the institute. The radio station was housed in the former magnetic observatory at Talstraße 35 from October 1925 onwards.

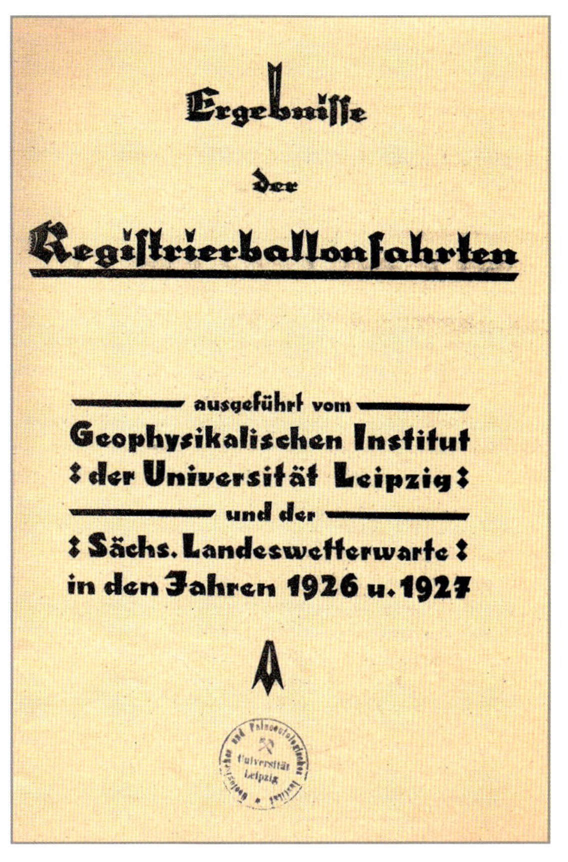

Ergebnisse der Registrierballonfahrten

ausgeführt vom

Geophysikalischen Institut der Universität Leipzig

und der

Sächs. Landeswetterwarte in den Jahren 1926 u. 1927

Figure 2.6: Front page of the issue on results from sounding ballooning.

As already mentioned in Chapter 1, starting from 1922, the institute's weather service also was made available to commercial and light aircraft taking off from the Leipzig-Mockau airfield. In the autumn of 1924, the Geophysical Institute even established its own flight weather station in Mockau. It continued until Halle-Leipzig airport was opened in Schkeuditz on July 1, 1927. The launch of the pilot balloons, which had so far taken place at 6 o'clock in the morning from the roof of the institute, were now made from the airfield. The series of double sightings of pilot balloons carried out by the institute – with support from the *Notgemeinschaft der deutschen Wissenschaft* (the forerunner of the DFG) – supplied valuable insights into wind structure in the lowest layers of the atmosphere, and the term "Leipzig wind profile" is used in technical literature even today. Data collection balloon launches were organized in cooperation with the *Landeswetterwarte* (district weather station) in Dresden between 1926 and 1932 under the direction of Paul Mildner; the results were published collectively (Fig. 2.6).

The study of the free atmosphere had been considerably promoted by what later became the International Aerological Commission of the International Meteorological Organization (predecessor of the World Meteorological Organization), itself established in 1896 by Hugo Hergesell. Under the presidency of Sir Napier Shaw, the Commission met in Leipzig from August 29 to September 3, 1927 (Fig. 2.7). The Geophysical Institute organized the meeting, which also

Comisión internacional para la exploración de la alta atmósfera. Leipzig 1927 (29 agosto - 3 septiembre)
1. Arctowski. – 2. Marczell. – 3. Lempfert. – 4. Miss Austin. – 5. Peppler. – 6. Doctora Lammer. – 7. Rinne. – 8. Zeissler. – 9. Hergesell. – 10. Moltschanoff. – 11. Bruhns. – 12. Hesselberg. – 13. Exner. – 14. Fontseré. – 15. Sra. Wallén. – 16. Sir G. Walker. – 17. Mariolopoulos – 18. Sir Napier Shaw. – 19. Richardson. – 20. Meseguer. – 21. Sra. Hesselberg. – 22. La Cour. – 23. Enge. – 24. Eredia. – 25. Bjerknes. – 26. Mildner. 27. Schmauss. – 28. Van Everdingen. – 29. Keil. – 30. Linke. – 31. Oishi. – 32. Weickmann. – 33. Hermann. – 34. Wallén. – 35. Roná. – 36. Cannegieter

Figure 2.7: The International Commission for the Study of the Upper Atmosphere (later renamed International Aerologic Commission) meets in Leipzig, August/September, 1927.

was attended by Bjerknes. The conference participants came to numerous practical agreements (e. g. concerning geopotential) and discussed many theoretical questions (such as solar radiation in the higher layers of the atmosphere).

Weickmann became president of the International Aerological Commission in 1935 and held this position until the end of the war. He had significant influence on the execution of scientific aerological flights. In a *synoptic-aerological investigation of the weather conditions during the international days of December 13–18, 1937*, carried out in collaboration with J. Bjerknes, P. Mildner and E. Palmén, Weickmann emphasized the importance of the international cooperation in the field of aerological research. In addition, he published important papers on aerological registration instruments, in particular radiosonde constructions, and on aerological diagram papers.

Weickmann supported all efforts at using an airship as a flying observatory for scientific research. Consequently, he was a member of "Aeroarctic" (International study association for the exploration of the Arctic by airship). The legendary arctic expedition by the airship "Graf Zeppelin" from July 24 –31, 1931, was certainly the highlight of these efforts (Fig. 2.8). The endeavour had been prepared intensively, for example with meteorological and aerological observations made by Weickmann during his cruise on the research and surveying vessel "Meteor" to Iceland and Greenland from July 16 to August 30, 1930. The "Graf Zeppelin" arctic expedition was aimed mainly at photogrammetric exploration of the area flown over in order to produce precise maps. The expedition itself was led by the famous airship pilot Hugo Eckener and supported by Notgemeinschaft der deutschen Wissenschaft, the Soviet government, the Ullstein publishing house and several private sponsors. Weickmann supervised meteorological observations and especially the multiple ascents made with the new radiosondes, which brought valuable results. The radiosondes had been developed by Moltschanoff, made fully functionally by Karolus, built by the Askania and Telefunken companies, and tested at Mockau airfield (Fig. 2.9). Plans for a North Pole expedition with LZ129 were abandoned after the "Hindenburg" crashed so disastrously at Lakehurst in 1937.

Figure 2.8: Ludwig F. Weickmann (on the left) and Hugo Eckener (on the right) in LZ 127 during the Arctic Expedition 1931.

In February 1926, Weickmann turned down a call to become director of the meteorological institute in Quito, Ecuador. In the following years, he also received a call to go to Hamburg to the German naval observatory; he decided, however, to stay in Leipzig. Rejecting this call allowed him to implement his long-held plans for an external observatory for seismic and earth magnetic research (see Chapter 9). He found a suitable location for the observatory on the Collm hill near Oschatz, some sixty kilometres from Leipzig. Despite the difficult economic conditions, the observatory was established in 1927,

Figure 2.9: Testing the radiosondes at Mockau airfield before the Arctic Expedition 1931.

Figure 2.10: Heinz Lettau, Paul Mildner, Vilhelm Bjerknes and Ludwig Weickmann at the Collm Geophysical Observatory in 1938.

with the construction taking five years. The official opening took place on October 6, 1932 (see Chapter 9).

Having the observatory meant that Weickmann could expand the spectrum of the institute from seismology (represented by Paul Mildner, Fig. 2.10) to include other fields involving the physics of the solid earth. For instance, he collected a series of measurements of all components of the earth's magnetic field over many years.

Heinz Lettau (Fig. 2.10) qualified as a professor with a habilitation thesis about a horizontal double pendulum for measuring the horizontal gradients in the Earth's gravity field. The proposal received considerable attention, not least from Werner Heisenberg. Lettau gained an international reputation with his book "Atmosphärische Turbulenz" (Fig. 2.11). Lettau also discovered an unusual magnetic concentration in Central Europe when surveying the Reudnitz earth magnetic anomaly near Collm. Weickmann was one of the first to supplement "Magnetische Reichsvermessung 1935.0" with regional and local measurements and geological interpretations. He was also one of the first to use geophysical measurements for practical purposes, in close co-operation with the head of the Saxon geological survey, Kurt Pietzsch (Fig. 2.12).

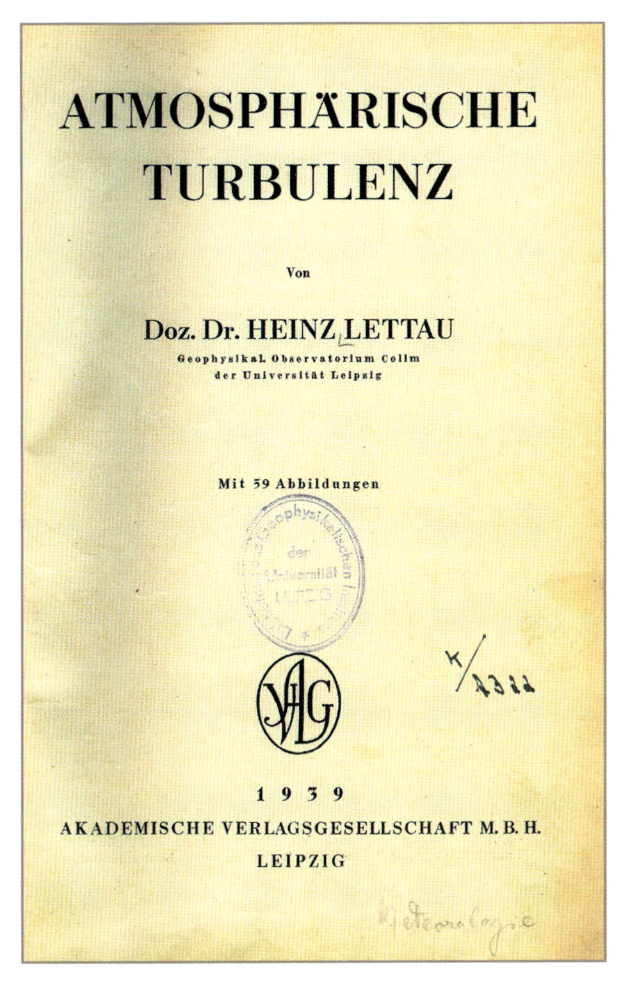

ATMOSPHÄRISCHE TURBULENZ

Von

Doz. Dr. HEINZ LETTAU

Geophysikal. Observatorium Collm
der Universität Leipzig

Mit 59 Abbildungen

1939

AKADEMISCHE VERLAGSGESELLSCHAFT M. B. H.

LEIPZIG

Figure 2.11: Title page of the book on atmospheric turbulence by Heinz Lettau.

Robert Lauterbach gained his a doctorate under Weickmann in 1938 with a thesis on geomagnetic investigations in Northwest Saxony and Northeast Thuringia. He would later become the leading geophysicist in Leipzig.

The further increase in the number of graduate students made it necessary to expand the institute on the third floor of Talstraße 38. I am deeply grateful to Professor Dr. Scheumann, director of the Mineralogical Institute and our "landlord", for his companionate and understanding household in respect to our increasing spatial claims. ... Altogether 66 people, including 4 women, earned a doctorate from the Geophysical Institute of the University of Leipzig. 70% of them turned to the national weather service ... The shortage of young people is gradually becoming apparent in this area as well, to such an extent that the students already have a job in their pocket, so to speak,

Figure 2.12: Ludwig F. Weickmann, Kurt Pietzsch and Finsch (from left) carrying out geophysical field measurements.

Figure 2.13: Leo Gburek in the Antarctic.

even before the end of their studies (Weickmann 1938, pp. 9f).

Staff members of the Geophysical Institute undertook several research expeditions. For example, Luise Lammert left for Australia for 16 months in 1928 after receiving a grant from the "International Federation of University Women". In Australia, she carried out radiation measurements with instruments made available by the "Notgemeinschaft der deutschen Wissenschaft". The expedition was also to investigate the application of front theory to the conditions of the Southern Hemisphere.

The Geophysical Institute under Weickmann endeavoured to maintain international cooperation even under the National Socialist government. A good example here is Erich Etienne's participation in the Greenland expedition run by the University of Oxford in summer/autumn 1938. Etienne set up his research programme for the expedition under Weickmann's guidance, undertaking geophysical measurements (radiation measurements, firn analysis), photogrammetric recordings and radio experiments. The Geophysical Institute provided a large part of the scientific equipment with the Collm Geophysical Observatory own precision engineering workshop producing the specialised instruments.

Leo Gburek, who also was involved in the earth magnetic survey of Saxony, represented the Geophysical Institute as the expedition geophysicist on the 1938–1939 German Antarctic expedition (Fig. 2.13). The roughly 2000-m Gburek Peaks near Schirmacher Oasis in Antarctica are named after Gburek, who failed to return from a weather reconnaissance flight over the North Atlantic in 1941.

The Jewish scientist Bernhard Haurwitz, one of Weickmann's numerous students and coworkers, received Weickmann's constant support despite the increasingly anti-Semitic atmosphere

Handbuch der Geophysik

Band III

Veränderungen der Erdkruste

Bearbeitet von

Prof. Dr. B. Gutenberg, Pasadena (Kalifornien)
Prof. Dr. F. von Wolff, Halle — Prof. Dr. A. Born †, Berlin
Prof. Dr. H. Heß, Nürnberg — Prof. Dr. E. Kraus, Berlin

Fortgeführt von

Dr. L. Weickmann
ord. Professor der Geophysik an der Universität Leipzig,
Direktor des Geophysikalischen Instituts

Mit 270 Abbildungen

Berlin
Verlag von Gebrüder Borntraeger
1940

Figure 2.14: The "Handbuch der Geophysik" from Beno Gutenberg (USA since 1930) was partly edited by Ludwig Weickmann.

in Germany. Together, they co-authored meteorological entries for Beno Gutenberg's *Manual of Geophysics*, which consisted of several volumes and was later edited by Weickmann (Fig. 2.14). At the end of 1932, Haurwitz went on a field trip to USA scheduled to last several months (Fig. 2.15). He did not return to Germany – not least because Weickmann advised him not to.

In the period from 1935 – 1945, the Geophysical Institute had to do without Weickmann's direction more frequently. By the middle of the 1930s, there were plans to unify the various weather services scattered around Germany into one national weather service, and Weickmann volunteered to serve in Berlin as provisional head of new *Reichswetterdienst* (national weather service) in the years 1935 and 1936. After the occupation of Norway in 1940, Weickmann was assigned the delicate task of directing the geophysical institutes there. At the

Figure 2.15:
Farewell party in 1932 for Bernhard Haurwitz (seated right) before his study trip to the US (next to Haurwitz is Ludwig Weickmann, standing in the middle is Luise Lammert).

Figure 2.16:
Participants of the *2. Studienaktion* on the tower platform of the Collm Geophysical Observatory. From left: Karlheinz Hinkelmann (later Professor at the University of Mainz), (slightly hidden) Herbert Käse (later Main Office of Climatology Potsdam), Ludwig A. Weickmann (later head of the analysis and forecast center of the Deutscher Wetterdienst), Ernst Lingelbach (later President of the Deutscher Wetterdienst), Horst Wiese (later head of the Geomagnetic Observatory Niemegk).

same time, he had to take over the organization of the weather service for the air force units in Norway and Finland. In early 1944, he was again drafted into the military, this time as an advisor on long-range forecasting for the air force operations staff based in Wildpark near Potsdam. Whilst Weickmann was fulfilling his military duties, Paul Mildner headed the institute and held meteorological and geophysical lectures.

During the war years, the Geophysical Institute was involved in two very differently structured programmes initiated by the *Reichswetterdienst*. The motivation for these training measures, so-called *Studienaktionen* or *Kurzaktionen*, was the great demand for meteorologists for the air force and the navy after 1939. The courses took place from spring 1940 to spring 1942 in Berlin, Hamburg and Leipzig (*1. Kurzaktion*) and from August 1941 to December 1943 in Berlin, Breslau, Danzig, Dresden, Leipzig, Prague and Vienna (*2. Kurzaktion*). Outstanding teachers in different fields were recruited for the training; in addition to Ludwig Weickmann, the names Albert Defant, Hans Ertel, Heinrich von Ficker, Friedrich Hund, Ludger Mintrop and Heinrich Schmitthenner must be mentioned. Well thought-out curricula and the ambition of the students guaranteed that the training was effective despite the brevity of the courses. The actions yielded 167 and 278 graduates respectively, who all received a diploma and thereafter saw service in quite different theatres. Although one in five did not survive the war, after 1945 many of the other "young meteorologists" found outstanding positions in weather service practice, research institutions, universities or foreign aid. They were based particularly in West Germany, but also in the GDR, in Austria and other countries (Fig. 2.16).

During the examination phase of the second *Kurzaktion,* Leipzig was hit by the heaviest air raid to date, on December 4, 1943, and the institute building was destroyed (Figures 2.17 and 2.18). The examination papers of the 86 can-

Figure 2.17: Destroyed institute. The remains of the meteorological instruments can be seen in the windows (cf. Figure 2.3).

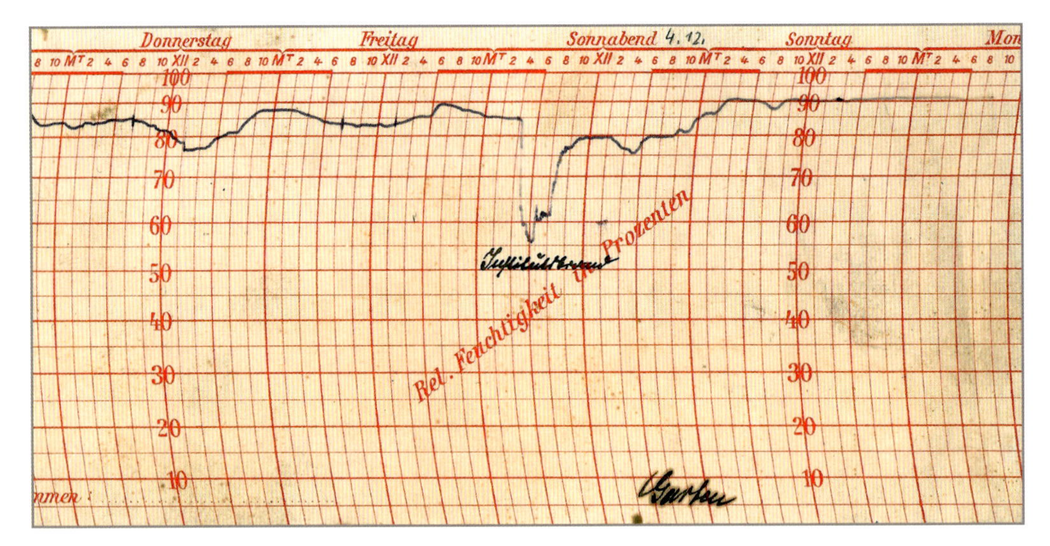

Figure 2.18: Humidity records; the sudden decrease is caused by the institute building burning on December 4, 1943.

didates were among the few things that could be saved, and so the examination could be ended correctly. With considerably foresight, a large part of the institute's extensive library had been taken to the Collm Geophysical Observatory and so was spared. After the bombing, the institute was assigned some rooms in the university-owned building at Talstraße 35. Fortunately, the observatory sustained no damage during the war; however, some losses occurred in the immediate post-war period, when it was abandoned by researchers.

The Geophysical Institute 1945–1955: Reconstruction and Stabilisation

Dietrich Sonntag, Peter Hupfer and Michael Börngen

Research life had gradually come to a standstill in Leipzig during the Second World War as many of the university's staff were drafted into military service. Air raids on Leipzig, particular in the night of December 3 to 4, 1943, destroyed about 80 % of the university buildings, including the University Astronomical Observatory and Talstraße 38, where the Geophysical Institute had its rooms.

Figure 3.1: Max Robitzsch (1887–1952).

Immediately after the war, the reconstruction of the institute began under the direction of Ludwig Weickmann. However, the encouraging rebirth was interrupted after only a few weeks when, on June 23, 1945 and shortly before the Red Army took control of Leipzig, the occupying Americans transported a number of eminent members of the university staff to Weilburg an der Lahn in the US occupation zone. Among them were the geologist Erich Krenkel, the mineralogist Karl Hermann Scheumann, the geographer Heinrich Schmitthenner, and the geophysicist and meteorologist Ludwig Weickmann and his research assistants Paul Mildner, Rudolf Penndorf and Dietrich Stranz. None of them would return to Leipzig. Other former assistants including Heinz Lettau were likewise no longer in Leipzig but continued their activity in West Germany and/or abroad. The result was that the Geophysical Institute was bereft of key personnel after mid-1945.

After Weickmann's retirement, Eduard Heise was commissioned with the provisional direction of institute business from end of June to November 15, 1945, before the university appointed first the physicist Friedrich Hund and later the mathematician Ernst Hölder as provisional directors of the Geophysical Institute and the Collm Geophysical Observatory; Hölder fulfilled the duties until the appointment of Max Robitzsch in 1950 (Fig. 3.1).

The Geophysical Institute was provisionally accommodated in the immediate neighbourhood of its old home, in Talstraße 35 (Fig. 3.2), a building that traditionally housed the geological and the mathematical institutes. The remaining core

Figure 3.2: Talstraße 35, domicile of the geological institutions. The Geophysical Institute was located here from 1943 bis 1951.

personnel, mainly the technical staff, were occupied with returning the rescued library stock to the shelves and putting the inventory into good order and condition. The institute's weather service began forecasting again from July 1945 on.

It was a fortunate circumstance for the institute that Weickmann's student Walter Hesse (Fig. 3.3) offered to work for the institute full time from August 1, 1945 (Hesse 1949). Working as research assistants were for a short time Alfred Mäde and from 1946 to 1953 Hildegard Nitsche. Heise and Mäde withdrew from the university in November 1945 and from that date on Hesse was charged with the direction of the institute by the Rector of the University of Leipzig.

Figure 3.3: Walter Hesse (1915–1979).

In August 1945, research activity recommenced at Collm Geophysical Observatory as well, and master precision engineer A. Schütz was tasked with replacing the lost equipment as far as possible. He was assisted by technical employees, among them E. Maasch, who had been on the institute's staff since 1918. They immediately started constructing a recording device for the magnetic control room and planning two magnetic variometers. They managed gradually either to procure most of the missing equipment or to produce it in the institute's own workshop.

The head of the Geophysical Institute was involved in the restructuring of the curricula and the study and examination regulations for meteorology and geophysics. The previously extensive slide collection used for teaching purposes had suffered losses, but it was restored and even significantly expanded. After the war (until 1949), 6 graduate meteorology students could complete their doctorates.

In January 1946, negotiations started regarding the reconstruction of the war-damaged weather radio house (the former earth magnetic observatory), and by late spring of the same year it was ready for occupation.

The University of Leipzig re-opened on February 5, 1946. Outstanding scientists judged as politically safe continued to work in the different faculties, while others returned from exile. New faculties opened (education, humanitarian sciences, financial/economic sciences and social sciences), as well as a so-called *Vorstudienanstalt* (VoSta) preparing children from proletarian backgrounds to study at the university. The VoSta in turn mutated into the *Arbeiter-und-Bauern-Fakultät*, the "workers' and peasants' faculty" (ABF), in 1949. The new faculties and the ABF were cadre schools training political officials for what later became the GDR. Taken together, the measures renewing the university and the new admission regulations for academic studies formed the *Erste Hochschulreform* (First University Reform).

Figure 3.4: Ernst Moritz Arndt; he directly converted the Morse signals into meteorological symbols.

Research at the Geophysical Institute and also at Collm Geophysical Observatory Collm gradually re-started from summer 1946 onwards after Hildegard Nitsche had taken over supervision of the observatory on July 1, 1946. She concentrated particularly on maintaining the seismic and the earth magnetic registration and evaluation.

Meteorological and geophysical lectures, practical courses and colloquia also began to be held at the institute after Hesse received a teaching assignment for meteorology and geophysics in 1946. Hesse was charged with running the institute under Hölder's direction after 1948.

The Soviets did not permit meteorologists to be trained at universities in the occupation zone until about 1947. The reason for this was probably that the *Reichswetterdienst* (national weather service) in Germany had been connected with the aviation ministry, and thus most meteorologists had been members of the air forces and so were under suspicion. As a consequence, students interested in meteorology registered to study geophysics and also graduated as *Diplom Geophysiker*. Until the early 1950s, the courses in meteorology and geophysics were much more closely linked than later on, and a large number of lectures were followed by students in both programmes.

At the beginning, the number of the students was small and included war veterans for the first six years after 1945. Some of the students especially interested in theoretical meteorology and geophysics moved to the University of Berlin, where the Institute for Meteorology and Geophysics had started training meteorologists and geophysicists under Hans Ertel.

In Leipzig, W. Hesse was the only person teaching courses in meteorology and geophysics until 1950, when Max Robitzsch was appointed. Teaching was really a Sisyphean task and it was therefore perhaps not surprising that quality suffered. By contrast, training in synoptic meteorology was good and included actual practice. Day by day data was collected via radio and drawn onto weather charts and topographies of the pressure field. E. M. Arndt (Fig. 3.4) and Cl. Auerbach were in charge of the technical tasks.

Only much later were teaching activities supported by hiring specialised instructors, usually experts from the meteorological service (Johannes Goldschmidt and Erwin Wiechert for general meteorology, Wolfgang Warmbt for climatology and bioclimatology, Josef Rink for aerology and Erich von Kilinski for atmospheric electricity) and from VEB Geophysik Leipzig (Gerhard Noßke for geoelectrics and Rudolf Meinhold for applied seismics).

The weight of the teaching load and the time devoted to saving and re-organizing the institute meant that research played only a minor role during this period. Similarly, the *Diplom* graduation theses from this time, which were written in parallel with the regular study programme, were less challenging than later ones.

The library, which was headed for many years by Melitta Werner, also suffered theft by the Soviet occupying forces. Nevertheless, it grew, saw heavy use and had money to purchase new publications. The international exchange of publications developed well, particularly after the institute began publishing again (in Akademie-Verlag Berlin) at the beginning of 1948.

The reinstatement of the Chair for Geophysics in 1950 clearly owed much to the engagement of Professor Hans Ertel, Director of the Institute for Meteorology and Geophysics at the University of Berlin. In a letter to the authorities for national education, Ertel argued forcefully in favour of the appointment of Max Robitzsch as Professor of Geophysics in Leipzig. Ertel's influence is evidenced in a letter dated April 4, 1949 to the Dean of the Natural Sciences Department within the Philosophical Faculty of the University of Leipzig from the Saxon State Government. It seems likely that Ertel argued to the authorities as he had in a letter to the philosophical faculty itself on August 22, 1949. In it, he wrote that during a visit to Sweden he had noticed the extraordinary interest in the Leipzig Chair for Geophysics among Swedish and other Scandinavian colleagues, and that they had interpreted the long vacancy of the chair as symptomatic of the decline of sciences in the Eastern Zone of Germany.

By this time, however, Ernst Hölder, the Dean of the Natural Sciences Department, had already reacted to the letter from the Saxon State Government from April. He had asked Ludwig Weickmann, the former chair-holder, for a reference on Max Robitzsch, and Weickmann had answered that Robitzsch was without doubt well-suited for a professorship in geophysics at the University of Leipzig. Hölder then asked Robitzsch whether he would be willing to accept the offer of the chair and to move to Leipzig. Robitzsch answered that he would be glad to come to Leipzig, *where, as young scientist, I was stimulated by professor Bjerknes, and later by my friend Robert Wenger and my colleague Weickmann* (PA 1159, sheet 12).

Robitzsch visited the provisional accommodation of the Geophysical Institute, presumably at the beginning of 1950, as he reported on the visit to the Ministry for National Education on January, 30, 1950: *The accommodations of the Geophysical Institute in Leipzig, Talstr. 35, are entirely insufficient for the work to be carried out there and also for lecturing. Three-and-a-half rooms are available at present, and these accommodate parts of the institute library (all other books are stored at Collm), the administration and the registry, the reckoner, draftsmen and the assistant Dr. Hesse, currently charged with the direction of the institute. For me as the future director, there is no room available. The library area presently also accommodates the trainees. [...] The scientific staff of the institute consists of Dr. Hesse, whose salary is paid from funds for the director's position, and a scientific auxiliary worker,*

Figure 3.5: Schillerstraße 6, domicile of the Geophysical Institute from 1951 to 1971.

Mr. Noßke, whose salary comes from the assistant position. The Collm Geophysical Observatory is only inadequately staffed by the research assistant Dr. Nitsche. He continued that a chief technician would be necessary and that the master precision engineer Schütz was insufficiently paid (Pa 1159, sheets 47–49).

At the beginning of summer term 1950, Robitzsch was appointed Professor and Chair of Geophysics, and also Director of the Geophysical Institute and the Collm Geophysical Observatory. At about the same time, the institute received reinforcement on the personnel front in the form of Dietrich Sonntag, who worked at the institute from 1950 to 1957, first as an auxiliary assistant, then as a research assistant and chief assistant. Beginning in mid-1955, he held the direction of the institute for one year (Chapter 4).

In 1951, the institute moved to the second floor of Schillerstraße 6 (Fig. 3.5), where it would remain until its closure in 1971. Only the weather radio station remained in the small annexe next to the main building in Talstraße 35 (Kortüm 1963, pp. 8–9).

The Second University Reform in East Germany in 1951 brought not only stricter regulations for the study programmes but also the introduction of Marxist-Leninist basic studies and mandatory instruction in Russian. From this point on, the students were organised into year groups and subjected to an obligatory timetable. The winter term of 1951 saw one group each of meteorology and geophysics students enrolling for the first time, each group comprising about 15 students.

The training of meteorologists and geophysicists in these years followed study plans which did not differ in principal from the curricula of today. The basics of mathematics and in experimental and theoretical physics were taught to a high standard, providing the students with the knowledge base necessary not only for understanding their own field of study but also for advancing it. Compulsory lectures also covered the development of newer philosophies and the relationship between natural sciences and society.

The focus of research during Robitzsch's directorship lay on:

- Preparation of the "Handbuch der Aerologie" (Manual of Aerology, Fig. 3.6)
- Final work on "Hann/Süring, Lehrbuch der Meteorologie", Volumes I & II
- Development of Assmann's aspiration psychrometers, hair hygrometers, and devices for the measurement of radiation of the sun and the sky
- Studies on the energy content of air masses

This phase in the development of the Geophysical Institute ended with the death of Robitzsch on June 4, 1952. It was to Robitzsch's credit that he had succeeded not only in saving the Geophysical Institute and the Geophysical Observatory, but also in putting them on a solid footing despite the difficult circumstances.

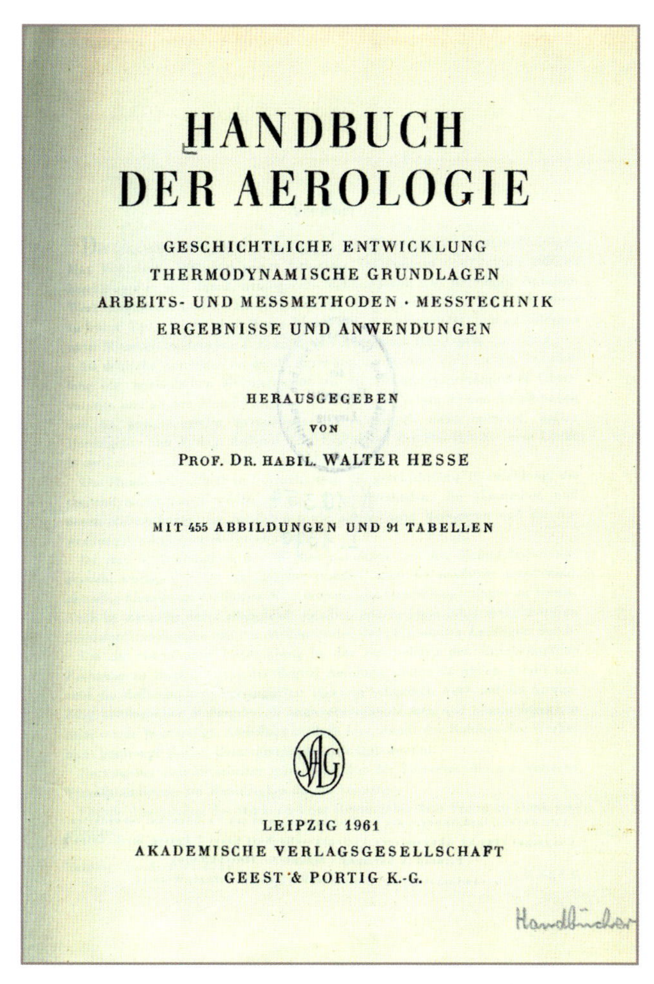
HANDBUCH
DER AEROLOGIE

GESCHICHTLICHE ENTWICKLUNG
THERMODYNAMISCHE GRUNDLAGEN
ARBEITS- UND MESSMETHODEN · MESSTECHNIK
ERGEBNISSE UND ANWENDUNGEN

HERAUSGEGEBEN
VON
PROF. DR. HABIL. WALTER HESSE

MIT 455 ABBILDUNGEN UND 91 TABELLEN

LEIPZIG 1961
AKADEMISCHE VERLAGSGESELLSCHAFT
GEEST & PORTIG K.-G.

Handbücher

Figure 3.6: Cover of the "Handbuch der Aerologie" (Manual of Aerology).

The previously broad geophysics department was split up into the more specialized disciplines of geophysics and meteorology in the spring term of 1953, when Hesse was a lecturer for meteorology and provisional institute director. Beginning in the autumn term, Hesse held a professorship and also had a teaching obligation in the agriculture faculty. In spring term 1954, Gerhard Fanselau, professor at Humboldt University Berlin and director of the Geomagnetic Institute Potsdam and the Geomagnetic Observatory Niemegk, came to Leipzig as a guest professor for geophysics (Kortüm 1963, p. 9).

4 The Geophysical Institute 1955–1971: Development and Closure

Peter Hupfer and Sigurd Schienbein

Recovery and New Perspectives: The Short Era Schneider-Carius

At the end of 1954, Karl Schneider-Carius (Fig. 4.1), until then employed by the German Meteorological Service (Deutscher Wetterdienst, DWD) in Bad Kissingen, appeared in the Geophysical Institute accompanied by his former colleague Hans Koch. Walter Hesse informed staff and students that Professor Schneider-Carius had taken over the directorship of the institute, and added that he would take over meteorology lectures by the end of the summer semester 1955. Thus the Hesse era at the institute came to an end.

After a short time at the Geophysical Institute, Schneider-Carius returned to Bad Kissingen to prepare to move permanently to Leipzig. Some slight difficulties arose firstly because the faculty knew nothing of his appointment, and secondly because negative assessments were written by Hans Ertel and Horst Philipps of Schneider-Carius, the State Secretariat's desired candidate for the chair originally held by Vilhelm Bjerknes.

The result was an interregnum of about one year – from mid-1955 to mid-1956 – when the institute was headed by Dietrich Sonntag. The period was marked by territorial squabbles and other unpleasantness instigated by Schneider-Carius from afar. Education nonetheless continued according to plan. In this context, it was a very positive development when Horst Philipps was appointed as successor to Hildegard Kohlsche to lecture in theoretical meteorology.

When the Chair of Geophysics was split to become the Chair of Meteorology and the Chair of General Geophysics (since 1958 Gerhard Fanselau, Figure 4.2), the difficulties with finding an appointee were solved: Schneider-Carius began work as Director of the institute and of the Collm Geophysical Observatory in May 1956.

The character, temper, experience and ideas of the new director introduced a sustained spirit of optimism at the institute. He promoted the inclusion of new fields of research – in addition to traditional seismology – at Collm Geophysi-

Figure 4.1: Karl Schneider-Carius (1896–1959).

Figure 4.2: Gerhard Rudolf Fanselau (1904–1982).

Figure 4.3: Erich Bruns (1900–1978).

cal Observatory by introducing ionospheric measurement and research into FM wave spread in the lower atmosphere. These last two fields had already been initiated by Alfred Adlung.

A part of the institute moved into the old Astronomical Observatory in Stephanstraße 3 in Leipzig. In the process, it found itself with a well-equipped workshop (headed by master mechanics Israel and after 1959 Werner Bohmann). An electronic workshop run by Ing. Günther Neubert was established at Collm Observatory to assist the research there, and practical training with meteorological instruments, run after autumn 1957 by Sigurd Schienbein, was also initiated there. Some time later, equipment for receiving weather data was also installed by Horst Schellenberg.

A major expansion of the institute resulted from the contact and interaction between Schneider-Carius and Erich Bruns (Fig. 4.3), head of the Hydro Meteorological Institute of the GDR's maritime hydrographic service SHD, who was looking for a partnership with an appropriate university-based institute. Bruns had obtained his postdoctoral degree (habilitation) at the Faculty of Mathematics and Natural Sciences and had since then given lectures in physical oceanography. Because oceanography was a fledgling science in the GDR and had its origins in the military, Bruns was always interested in establishing a larger department of oceanography at the university and leading such a department himself. Ideas and visions like this were much to Schneider-Carius' taste because they contributed to bringing his vision of uniting the three main branches of geophysics at one institute within realistic reach. However, about two years later Bruns and his SHD were offered the opportunity of becoming part of the (East) German Academy of Sciences as the Institute of Oceanography, an opportunity that he grabbed with both hands. Nonetheless, the cooperation between Schnei-

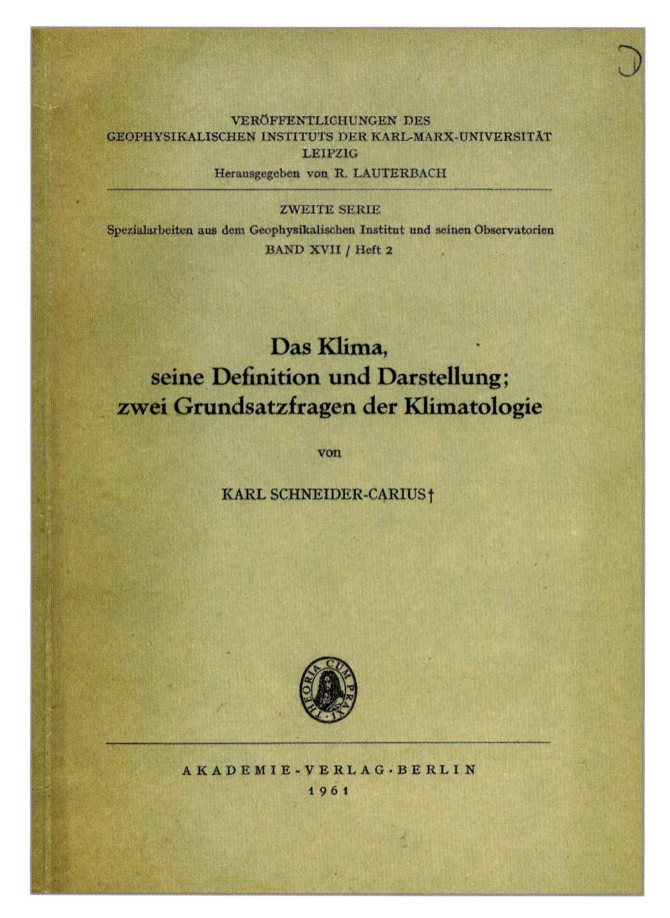
VERÖFFENTLICHUNGEN DES
GEOPHYSIKALISCHEN INSTITUTS DER KARL-MARX-UNIVERSITÄT
LEIPZIG
Herausgegeben von R. LAUTERBACH

ZWEITE SERIE
Spezialarbeiten aus dem Geophysikalischen Institut und seinen Observatorien
BAND XVII / Heft 2

**Das Klima,
seine Definition und Darstellung;
zwei Grundsatzfragen der Klimatologie**

von

KARL SCHNEIDER-CARIUS†

AKADEMIE-VERLAG·BERLIN
1961

Figure 4.4: Much attention found the booklet on some basis issues of climatology by Schneider-Carius.

der-Carius and Bruns resulted in the establishment of the Maritime Observatory of the Geophysical Institute in Zingst on the Baltic coast (see Chapter 10).

Schneider-Carius himself dealt with fundamental issues of climatology. He planned a monograph on trade winds and monsoons, which unfortunately remained uncompleted. In a paper published in the scientific journal of what was then called Karl-Marx-Universität, he critically discussed the state of the weather forecast and contemporary reception of Vilhelm Bjerknes' ideas on the matter (Fig. 4.4). He also published other studies, including papers on the history of meteorology.

Meteorology studies continued as before, with guest lecturers bringing outside views to the institute. A new element was added in the form of weather flight training from Kanitz airfield near Riesa in collaboration with the Aviation Faculty at Dresden Technical University. Sadly, however, problems in Dresden resulted in the closure of the programme after only a few years. There was also some discussion of setting up regular trips for Leipzig students to Austria to measure glaciers, which was feasible at that time despite the Cold War.

An early highlight in the research activities was the institute's participation in the International Geophysical Year and related programmes in 1957/59 (at the Collm Observatory, participation by Leipzig researchers in Atlantic expeditions on board the research vessel M. Lomonosov, and other activities). In addition, many research liaisons developed, especially with scientists in the Soviet Union (Fig. 4.5) and eastern European countries. For example, some years later (1963–1964), Peter Nitzschke was a member of the 8. Soviet Antarctic Expedition. His results were published in the institute's publication series (Nitzschke 1966; Fig. 4.6).

Figure 4.5: Schneider-Carius (center) with the Russian scientists Sverev (left) and Katschurin (right) (at the end of the 1950s).

There was something of an éclat when Dietrich Sonntag suddenly resigned from the institute due to personal differences with Schneider-Carius. Sonntag had been responsible for a sizeable part of the teaching and research at the institute, and the situation was deteriorated when his long-standing assistant Friedhold Weber left in 1958. In compensation, Christian Hänsel (physics of the atmosphere, synoptic meteorology), Peter Hupfer (oceanography and maritime meteorology) and Sigurd Schienbein (meteorological measurement methods) were hired as new research assistants at the beginning of winter semester 1957. Karl-Heinz Bernhardt joined at the same time as a doctoral candidate with an interest focusing on theoretical meteorology and the physics of the atmospheric boundary layer. The map which Schneider-Carius laid out for his young team had a seminal effect on them and their successors for decades, its influence enduring even until today in some cases. For instance, Rudolf Schminder was employed as a research assistant in 1959, later heading the Collm Geophysical Observatory until the end of the century and turning it into a widely acknowledged centre of research excellence.

Overall, the institute grew both quantitatively and qualitatively thanks to Schneider-Carius' activities. Numerous doctoral and habilitation theses were completed, cooperation with domestic and foreign institutions increased. The colloquia organized in the Scheider-Carius years were legendary, featuring as they did lectures by many prominent German and international meteorolo-

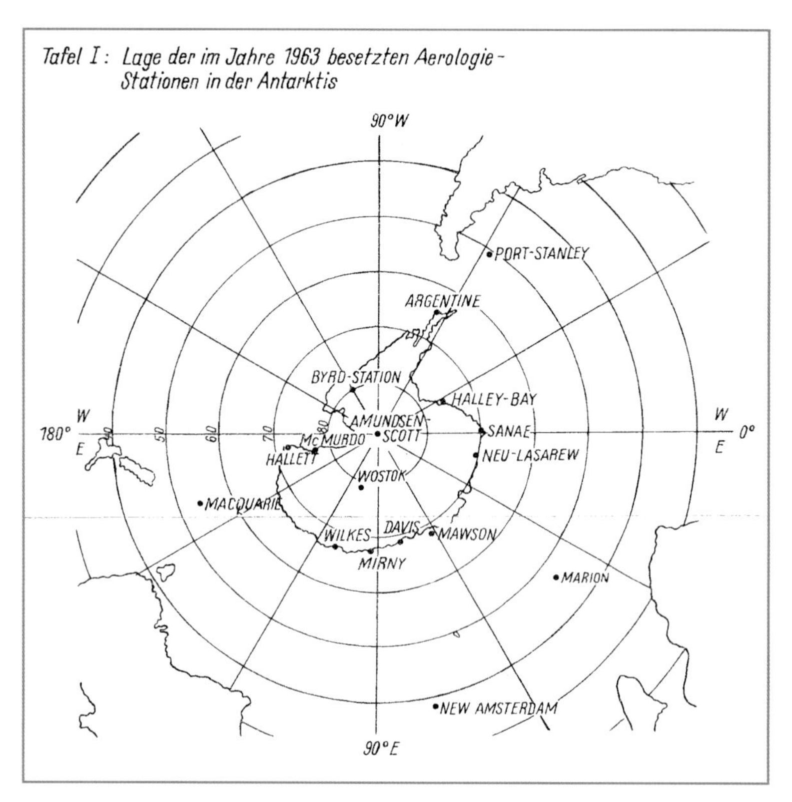

Figure 4.6: Peter Nitzschke published in 1966 results of his meteorological studies in the South Polar region. The publication contains a map of aerology stations in 1963.

gists. A particular highlight was the glamorous institute conference held as part of the celebrations for the 550th anniversary of the University of Leipzig in autumn 1959.

The sudden death of Schneider-Carius on December 1, 1959 in front of Leipzig's main station deeply shocked the institute's staff and students alike. The faculty decided that the geophysicist Robert Lauterbach would temporarily take over the directorship, while the changes initiated by Schneider-Carius were already stable enough to continue and even be expanded on.

The Geophysical Institute conducted by Friedrich Kortüm

In early 1960, Horst Philipps, Director of the Meteorological Service of the GDR, explained to the research personnel at the institute that he proposed the appointment of Friedrich Kortüm (Fig. 4.7), a lecturer in forest meteorology at the Forestry Faculty of Humboldt University Berlin in Eberswalde. Kortüm was a versatile meteorologist who had already proved himself in various fields of activity, as well as being director of the main Office for Climatology of the Meteorological Service. A man of considerable integrity, Kortüm was predestined to manage the institute, which had become ever larger and was by no means easy to administer. He was appointed as Acting Director of the institute and its observatories on January 1, 1961. Despite being less than satisfactory for the institute, Kortüm's appointment could only be as acting director and professor as he had not completed his postdoctoral habilitation at the time. Philipps himself took over the Chair of Meteorology on a part-time acting basis. Bruns had been appointed professor with teaching responsibilities in 1960, also on a part-time basis. As Director, Kortüm had little time to complete his habilitation at Humboldt University and so it was only in 1966 that he finally took over the directorship of the institute as a full professor.

Figure 4.7: Friedrich Kortüm (1912–1993).

The tradition of employing teaching staff from the Meteorological Service was continued by contracting Gerhard Dietze to give courses in atmospheric optics and Hans Hinzpeter for radiation (until 1961). Weickmann's former student William Peine conducted mathematics exercises with the students. Kortüm himself supervised numerous doctoral dissertations.

In addition to the work done at the Collm Geophysical Observatory and the Zingst Maritime Observatory described elsewhere in this volume (Chapters 9 and 10, respectively), a number of research projects were also pursued at the institute itself in Leipzig, with the majority coming under the general framework of heat balance.

The projects focused on the heat balance of the Earth's surface on the North German Plain (Kortüm), atmospheric heat flows with particular emphasis on boundary layer processes (Bernhardt), heat transport in the atmosphere from a synoptic viewpoint (Hänsel), and heat balance and temperature conditions in the near-shore zone of the Baltic Sea (Hupfer). Relatively extensive studies brought new insights into the long-term behaviour of meteorological and oceanographic parameters in the western Baltic Sea. The results led to investigations of climate impact on the coast (Hupfer).

Industrial Meteorology

As from 1958 the new sub-discipline of Industrial Meteorology was being developed at the institute (Koch, Schienbein), much to the consternation of the Meteorological Service of the GDR. Industrial Meteorology involved applying meteorological principles and measurement methods to industrial processes and so links the physics of the atmosphere with technical air treatment.

After taking over as Director, Schneider-Carius soon proposed reviving the previous cooperation with industry within the framework of applied meteorology, and charged Hans Koch with furthering the idea. Related fields in agriculture and medicine were already under investigation under the headings Agricultural Meteorology and Biometeorology.

In response to requests from the rubber industry, rubber samples were exposed to the atmosphere and evaluated. As a result, a central natural weathering station for the rubber industry was installed at the Collm Observatory providing observations on the effects of air temperature, humidity, wind velocity and direction, global and sky radiation, UVA and UVB radiation, and ground-level ozone. The facility meant that the normal weathering processes and ageing of rubber could be observed and evaluated in the light of scientifically recorded meteorological parameters.

Figure 4.8: Front cover of the book "Industrie-meteorologie".

These first investigations soon prompted inquiries from the lignite mining and textile industries, as these were also activities in which the air condition has a major influence on the properties and processability of materials. For the textiles industry, air movement and air condition are of evident importance for the quality of the end products during the production of high polymer threads. In the lignite (or brown coal) mining industry, the efficiency of the raw coal drying process is also obviously affected by meteorological conditions. Taking into account the outdoor weather, drying could be significantly improved by a targeted design of ventilation. The list of similar inquiries could be extended, showing that there was considerable interest in meteorology-related problems in various fields of industry. Hans Koch published "Industriemeteorologie" in 1969 (Fig. 4.8), discussing some of the projects being undertaken at the time, some as *Diplom* theses and doctoral dissertations. In addition to Koch and Schienbein themselves, the group expanding this new field also included Alfred and Sabine Helbig, Ulrich Müller, Peter Nitzschke and Eberhard v. Schönermark.

The attention that the activities of the Industrial Meteorology working group received inside and outside the university was generally positive. This culminated in the University's Rector granting permission to establish a dedicated department at the institute on January 1st, 1968, at a time when the closure of the institute had, in fact, already been decided.

At the root of the intense clashes with the Meteorological Service was the controversial question of whether the issues dealt with by the industrial meteorology work group were still within the scope of meteorology or not.

The Closure of the Geophysical Institute in Leipzig

Hupfer (2012) relates that rumours began to emerge in early December 1965 that the Geophysical Institute would be closed as part of the Third University Reform in East Germany. The authorities decreed that education of meteorologists should be concentrated at Humboldt University in Berlin, and the establishment of a "Central Meteorological Education Institute" was discussed, although this last never actually happened. After the Ministry of University Education had made its decision, it emerged that there were several well-founded objections to closing the Geophysical Institute, chief among them being the size difference, the practical relevance of the education and also the central location of the institute. The fact that these objections played no role whatsoever leads to the assumption that the state Meteorological Service, with its very pronounced agenda of centralisation, was the driving force behind the perplexing decision. The passive resistance mounted by almost all members of the institute staff meant that the observatories at Collm and Zingst were not transferred as planned, to non-university institutions, but remained parts of the University of Leipzig. Similarly, the workshops, the instrumentation, and most of the library remained in Leipzig and were further developed under the auspices of the Department of Geophysics (see Chapter 6). Of the permanent research staff, only K.-H. Bernhardt (1968) and F. Kortüm (1971) continued their work, based at the Department of Meteorology (within the Section of Physics) in Berlin-Friedrichshafen. The meteorologists of the Industrial Meteorology group were given the opportunity to establish a "Research and Information Centre for Special Production Optimization" under the umbrella of *VEB Kombinat Wolle und Seide* (state-held wool and silk combine). The new name was indicative: it deliberately omitted any reference to meteorology, meaning that the "Centre" did not come into conflict with state Meteorological Service. The Research and Information Centre continued very successfully with about 20 employees until 1992.

All students of the register Meteorology 1966 finished their study according to the plan in January 1971. Thereafter the Geophysical Institute in Leipzig was closed. The rooms in Schillerstraße 6 had to be abandoned by the middle of the year. The remaining staff, the library and other remnants were transferred to the new Department of Geophysics at Talstraße 35.

Finally, and perhaps inspired by the spirit of upheaval in autumn 1989, in the last months of the GDR's existence several petitions were submitted to the Rector of the University of Leipzig requesting permission to re-establish a Department of Meteorology and so reverse the unwarranted decision of 1965/1966.

The Institute for Geophysical Investigation and Geology 1958–1969

Franz Jacobs

In the early 1950s, a working group on applied geophysics was established at the Geological-Palaeontological Institute of the University of Leipzig under Robert Lauterbach (Fig. 5.1). The development and application of geophysical methods to geological investigation was the focus of teaching and research, particularly with regard to exploring for natural mineral deposits, energy resources and groundwater. Finally, on March 1, 1958, the Institute for Geophysical Investigation was founded *despite substantial opposition from Freiberg*, as Lauterbach liked to remark. In 1965, the institute expanded by incorporating the Geological-Palaeontological Institute, and was renamed, becoming the Institute for Geophysical Investigation and Geology (Institut für Geophysikalische Erkundung und Geologie).

Figure 5.1: Robert Lauterbach (1915–1995).

Lauterbach had been successful in attracting increasing numbers of students. In addition, the Geophysical Institute had closed its *Diplom* programme for geophysicists, and Lauterbach and his colleagues had taken over. Thus they filled a gap in the spectrum of scientific education at the university, a gap that was problematic in view of the rapidly growing demand for geophysicists on the job market. Between 1954 and 1972, about ten to fifteen students each year graduated as *Diplom Geophysiker* from the chair for Applied Geophysics under Lauterbach in the Faculty of Mathematics and Natural Sciences (Fig. 5.2). He was greatly assisted by Magdalena Meissner and Walter Neumann (Fig. 5.3). There were also several dozens off geophysical engineering graduates in the 1960s.

From the beginning, life at the institute was coloured by the substantial advantage of being located close to *VEB Geophysik Leipzig*, the state-owned company that emerged at the beginning of the 1950s from the *Geophysikalischer Dienst* (Geophysical Service) of the *Deutsche Geologische Landesanstalt* (National Geological Institution). VEB Geophysik Leipzig's task was to make use of the full range of geophysical techniques to identify and secure supplies of natural

Figure 5.2: Students of Geophysics in front of the building Talstraße 35 in the fifties (montage).

Figure 5.3: "Schatzsucher unserer Zeit", W. Neumann, Verlag Enzyklopädie Leipzig, 1961.

resources for the German Democratic Republic, and it was accordingly generously equipped. For example, the staff grew from 500 to 3000 between 1955 and 1970, and most Leipzig graduates in geophysics found their first job at VEB Geophysik. The institute itself profited substantially from access to external instructors and by being able to share geophysical equipment. For students, the practical courses were exciting and the topics for graduation theses were interesting, and so the cooperation was popular. From the beginning, Robert Lauterbach himself also acted as research director at VEB Geophysik. In addition, he oversaw the development of the "Geophysikalische Karte der Deutschen Demokratischen Republik" as its editor.

The close co-operation between university and industrial geophysics was also reflected in the joint development and practical introduction of new geophysical procedures:

- Geomagnetics (micromagnetics, proton magnetometer, isanomal statistics) (Figures 5.4, 5.5, 5.6)

- Radon emanometry, gamma-ray spectrometry, radium metallometry (Fig. 5.7)
- Geophysical procedures in lignite mining and archaeology (Fig. 5.8)
- Applied seismics (analog and digital reflection seismics, group shots, multiple reflections, shear waves, vertical profiling) (Figures 5.9, 5.10)

Figure 5.4:
Magnetic field balances according to Fanselau.

Figure 5.5:
Protonmagnetometer for Aeromagnetics. Aeroport Leipzig-Mockau (around 1960).

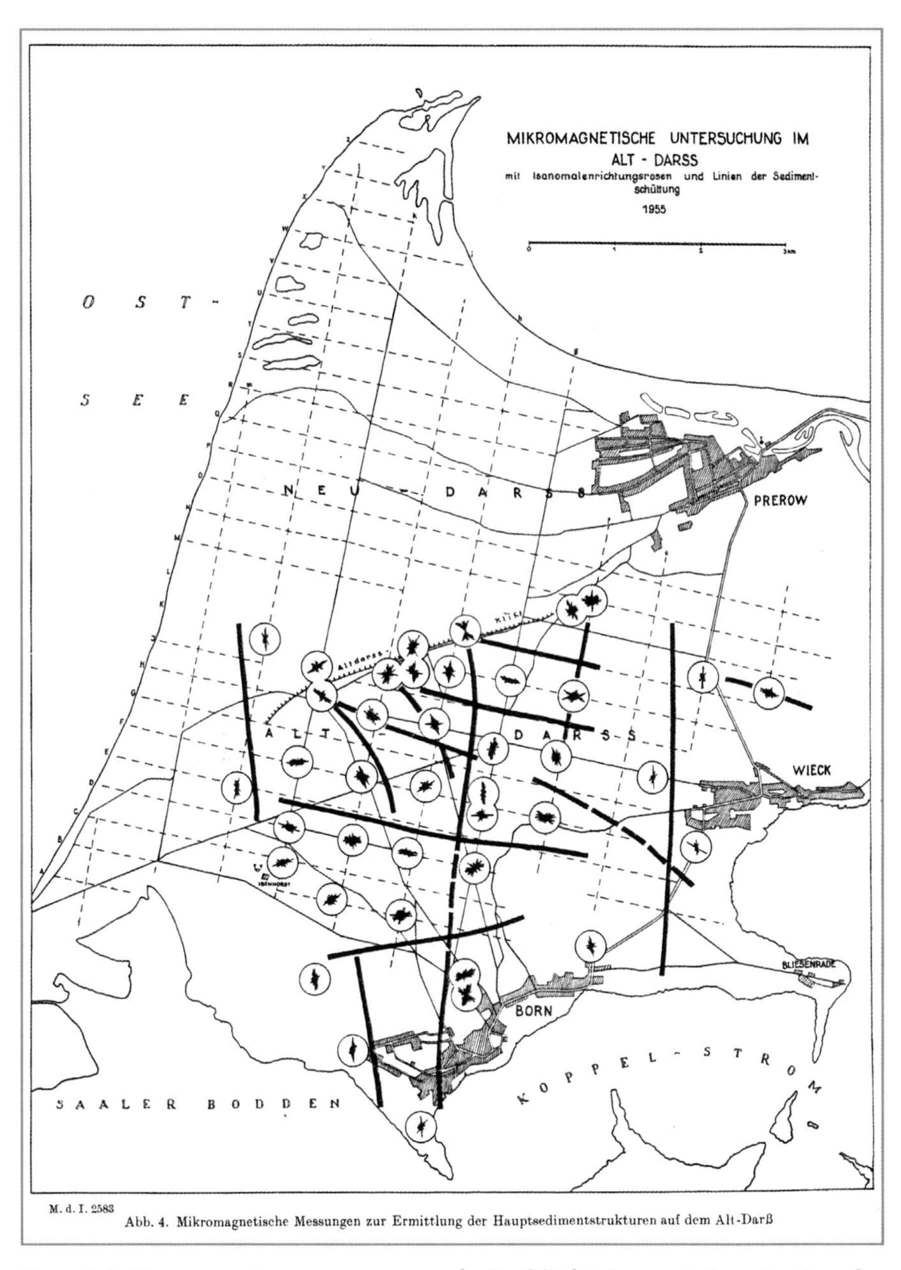

Abb. 4. Mikromagnetische Messungen zur Ermittlung der Hauptsedimentstrukturen auf dem Alt-Darß

Figure 5.6: Micromagnetic measurements on the Darß/Baltic Sea coast. Investigation of sediment structures. 1955.

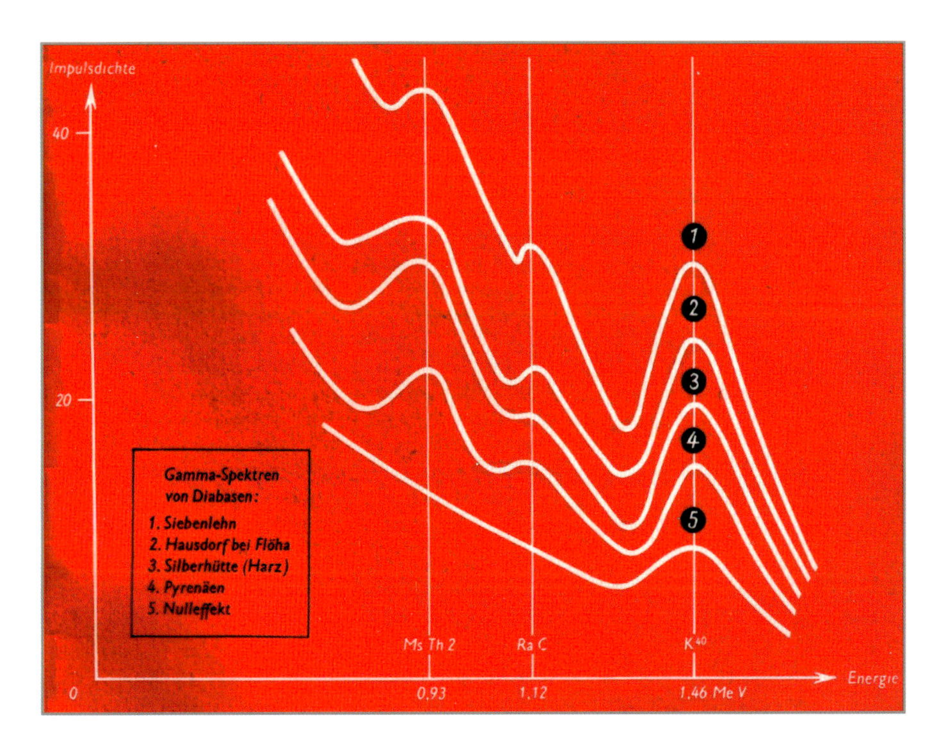

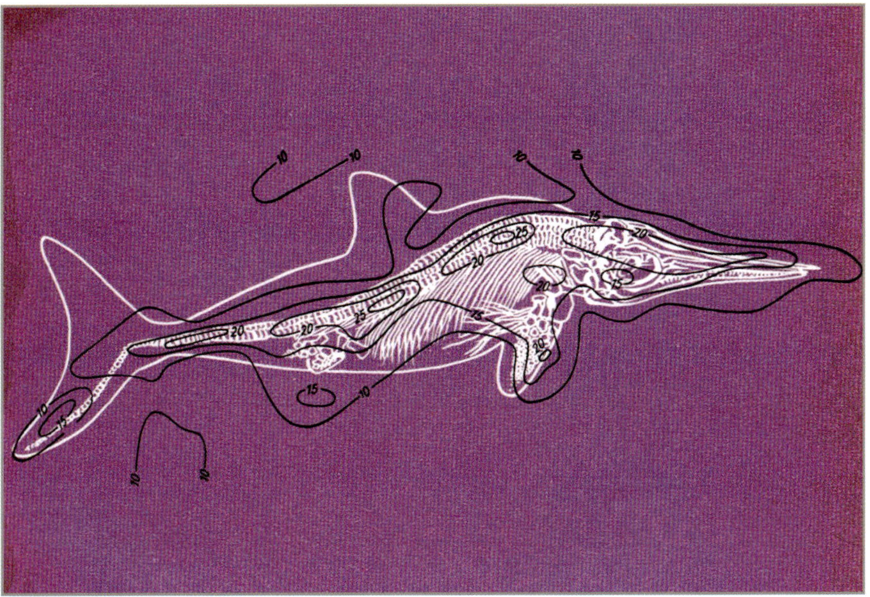

Figure 5.7: Gamma-Spectrometry. Top: Content of uranium, thorium, potassium in rock samples, 1964. Bottom: Radioactivity (Impulse density) of an Ichthyosaurus. Geological-Palaeontological Collection, 1962.

Figure 5.8: Geoelectrical resistivity measurements for investigation of stability of lignite open mining large equipments.

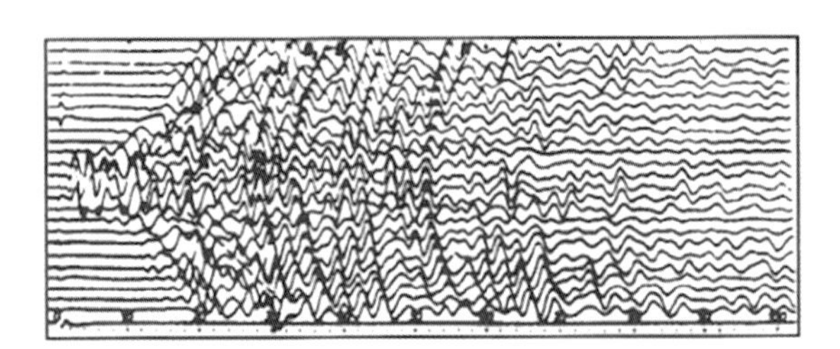

Figure 5.9: Shear wave seismics 1968. Seismogram for exploration of lignite seams (by Grässl and Patzer).

Figure 5.10: Steffen Grässl (1941–1984).

Numerous research results were published in *GEOPHYSIK UND GEOLOGIE – Beiträge zur Synthese zweier Wissenschaften,* the journal published by B. G. Teubner Verlagsgesellschaft, Leipzig, between 1959 and 1970. Its 15 volumes were the continuation of geoscientific publications which had begun within *Wissenschaftliche Zeitschrift der Universität Leipzig, Mathematisch-Naturwissenschaftliche Reihe* from 1953 to 1956 (Fig. 5.11).

Figure 5.11: Journal "Geophysik und Geologie", Volume 9. Verlag B. G. Teubner Leipzig, 1966.

The Institute for Geophysical Investigation organized the 22nd Annual Conference of the German Geophysical Society (Deutsche Geophysikalische Gesellschaft DGG e. V.) from May 2–5, 1958 in Leipzig, the place the society had been founded. Aside from chairing the conference, Lauterbach also took over the vice presidency of the society. 246 guests (153 from East Germany, 65 from West Germany, 24 from other countries) participated in 51 lectures and numerous excursions. This would remain the only DGG Annual Conference to take place in East Germany (Fig. 5.12).

Figure 5.12: 22nd Annual Conference of the German Geophysical Society in Leipzig 1958. Program (Cover).

6 The Department of Geophysics in the Section of Physics 1969–1993

Franz Jacobs

With the Third University Reform in the GDR of 1968, the traditional institutes, including the Institute for Geophysical Investigation and Geology and the Geophysical Institute, were closed in 1969 and 1971, respectively. The associated courses in geophysics and meteorology at the University of Leipzig were also closed down. The directors of the institutes were released from their responsibilities, a sorry act of political arbitrariness in the case of Robert Lauterbach, while the relocation of academic education in geophysics to Freiberg and in meteorology to Berlin had particularly negative consequences for the University of Leipzig. It became very obvious that there was a hole in the education offered by the university because the general interest in earth sciences was particularly pronounced in the region around Leipzig as a direct result of the many traditions in the area and also of the current demand.

Thanks to the efforts of Robert Lauterbach and physicists Artur Lösche and Werner Holzmüller, most of the staff and working groups in the Institute for Geophysical Investigation and Geology were transferred to the Department of Geophysics (Fachbereich Geophysik, later Wissenschaftsbereich Geophysik) of the Section of Physics when it opened in January 1969. For the Geophysical Institute, however, the departure of prominent researchers to Humboldt University in Berlin was a great loss; parts of the institute, including both observatories and the technical workshop with its staff, remained at the University of Leipzig and were integrated into the Department of Geophysics.

The department retained its home at Talstraße 35 and Stephanstraße 3, and also the premises of the Collm Geophysical Observatory near Oschatz and the Zingst Maritime Observatory on the Darß peninsular on the Baltic coast. However, the Geophysical Institute rooms in Schillerstraße 6 had to be given up. New professorships were created in addition to Robert Lauterbach's: Geonomy (1970, Gerd Olszak), Environmental Geophysics (1979, Christian Hänsel) and Applied Geophysics (1980, Hans Rische). This allowed the spectrum of research in geophysics, meteorology and oceanology to be safeguarded and even expanded, despite the lack of input from students' work.

Figure 6.1: VEB Geophysik Leipzig (the so-called "Industriepartner"). Seismic shot group in the seventies.

In geophysics, research was characterized in terms of content by applied geophysical topics, in terms of technology by the rapid development of digital data acquisition and processing, and in terms of organization by the proximity to VEB Geophysik (Fig. 6.1) and its expansion into other countries in the Eastern Bloc. The main topics of investigation were:

- Digital seismics / Deep seismic soundings
- Signal processing / Spectral analysis
- Modelling and inversion
- Lithological analysis and statistical interpretation
- Radioactivity, petrophysics and biogeophysics (Fig. 6.2)

The Department of Geophysics achieved an international reputation in the area of applied seismics thanks to the efforts of Steffen Grässl and to industrial contracts during the introduction of digital reflection seismic surveys at VEB Geophysik Leipzig.

The CLL seismological station at Collm Geophysical Observatory – headed by Bernd Tittel from 1965 – also enhanced its international reputation in global earthquake observation (Fig. 6.3).

Tough negotiations brought about special arrangements allowing academic education to continue in some form despite the restructuring. Scientists from Leipzig, geophysics students from Freiberg and some foreign students, mostly

Figure 6.2: "Geophysik und Umwelt", C. Hänsel, Urania-Verlag, 1975.

from Africa, completed their degrees as *Diplom Geophysiker* in Leipzig. In meteorology, academic education had to be restricted to supporting *Diplom Physiker* and *Diplom Lehrer* in mathematics and physics, whereby the meteorological and oceanological courses in Zingst enjoyed great popularity.

Robert Lauterbach had managed to have the Department of Geophysics designated by national decree as the GDR's Weiterbildungszentrum Geowissenschaften (Postgraduate Education Centre for the Geosciences) in 1970, an achievement that can be seen as an acknowledgment of its research competence. Postgraduate training for the so-called geoscientific cadres was intended to guarantee the efficiency of nationalised natural resource suppliers providing the country with energy. Supported by the Ministry of Geology and many state-owned companies and institutions, approximately 5000 experts participated in several hundred special one-week courses at the centre. They were introduced to the latest findings and tendencies in earth sciences

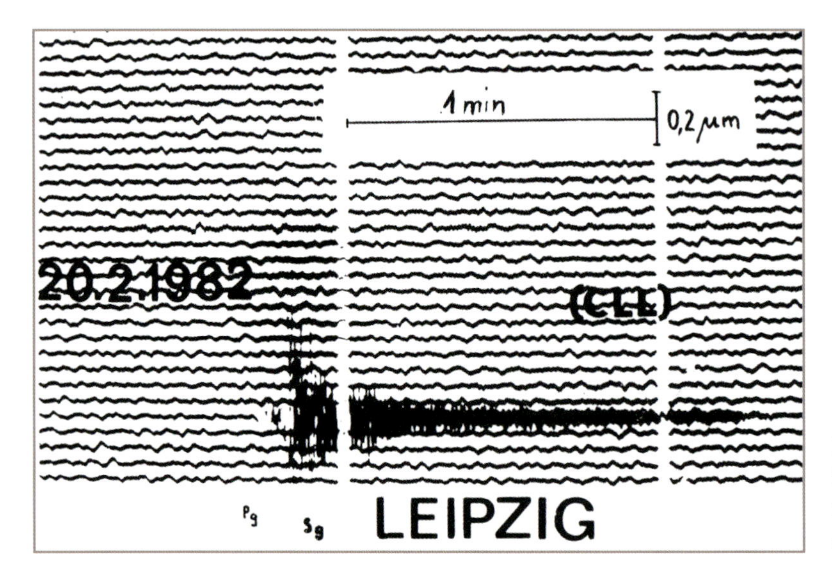

Figure 6.3: Leipzig Earthquake, 20.2.1982, Collm Observatory.

KARL-MARX-UNIVERSITÄT LEIPZIG
SEKTION PHYSIK
FACHBEREICH GEOPHYSIK

Leipzig, den 20. September 1971

Programm für den im Rahmen der postgradualen Weiterbildung vom 8.–12. Nov. 1971 im Hörsaal des Fachbereiches Geophysik, 701 Leipzig, Talstraße 35, stattfindenden Lehrgang

„Fortschritte und Entwicklungstendenzen der Geologischen Wissenschaften I"

Montag, den 8. November 1971

9.00–10.00 Prof. Dr. R. LAUTERBACH
Die Entstehung und Entwicklung des Lebens auf der Erde aus geophysikalischer Sicht

10.00–12.00 Prof. Dr. H.-J. TREDER
Kosmo- und Geogenese

KARL-MARX-UNIVERSITÄT LEIPZIG
SEKTION PHYSIK
WISSENSCHAFTSBEREICH GEOPHYSIK

Talstraße 35, Leipzig, 7010
Februar 1990

PROGRAMM

für den Lehrgang

„Fortschritte und Entwicklungstendenzen der Geologischen Wissenschaften" – Stand 1990

vom 2. bis 6. April 1990

Lehrgangsleitung: Prof. Dr. R. LAUTERBACH
Dr. sc. F. JACOBS

Lehrgangsort: Talstraße 35, Leipzig, 7010
1. Etage, Hörsaal 17

Figure 6.4:
Geoscience Postgradual Education Center. Program (details) of first /last week course 1971/1990.

Figure 6.5: "Physik des Planeten Erde", R. Lauterbach, Akademie-Verlag , Berlin 1985.

Figure 6.6: "Der Mensch und die Planeten", R. Lauterbach, Urania-Verlag Leipzig-Jena-Berlin 1987.

Figure 6.7: "Signale aus der Erde", F. Jacobs/H. Meyer, Verlag B. G. Teubner, Leipzig 1992.

and acquired the ability to apply this knowledge in their professional practice (Fig. 6.4). About 30% of the courses were devoted to geophysical topics. In addition, more than twenty geophysics engineers attended courses leading to a *Diplom* degree in two-week blocks between 1972 to 1976.

In the 1970s and 1980s, the Department of Geophysics published numerous scientific and popular-scientific treatises (Figures 6.5, 6.6, 6.7). Some were also published under licence in West Germany. *GEOPHYSIK UND GEOLOGIE, Geophysikalische Veröffentlichungen der Karl-Marx-Universität Leipzig, Dritte Serie* was published by Akademie-Verlag Berlin, combining and continuing the two earlier journals issued by the two institutes, now closed. A total of 16 volumes appeared from 1974 to 1992, dedicated to various topics in the Earth system from the deep interior of the Earth, through the Earth's crust, and up to the physics of the atmosphere and hydrosphere (Fig. 6.8).

Figure 6.8: "Geophysik und Geologie", Volume I/1 Meeresforschung, Akademie-Verlag, Berlin 1974.

In meteorology, research continued apace and in fact developed further at the Collm Geophysical Observatory – under Rudolf Schminder – with wind system measurements in the upper atmosphere, and in Zingst – under Peter Hupfer, in the 1980s under Hans Jürgen Schönfeldt – with research on the oceanographic and meteorological features of coastal areas. In Leipzig itself, research began on atmospheric environmental issues, particularly by Christian Hänsel (Fig. 6.9). Special mention should be made of spectral radiation measurements of dust and aerosol regulation and of the heat balance of the city atmosphere in this context. A *Stadtklimatologisches Observatorium* was set up in the university's high-rise building in the city centre and equipped with a laser measuring device linked with the former astronomical observatory in Stephanstraße 3. New instruments were developed and tested in the technical workshop, especially recording devices for the coastal range: buoys with oceanographic and meteorological sensors, and underwater acoustic equipment for analysing water bodies. However, it was only marginally possible to pursue the main objective of the Geophysical Institute as set out at is foundation in 1913 – namely basic research in meteorology.

The political changes in East Germany beginning in October 1989 and the subsequent reunification of Germany made it possible for geophysics and me-

Figure 6.9: Christian Hänsel (1924–2011).

teorology to make a new start at the University of Leipzig. As early as May 1990, the board of the Section of Physics agreed to re-establish basic study courses in geophysics and meteorology, and two years later *Diplom* programmes in geophysics and meteorology welcomed their first students.

On June 25, 1990, the newly-created *Kollegium zur Förderung der Geowissenschaften an der Universität Leipzig* drew up a programmatic memorandum on re-organizing studies and research at the earth sciences institutes, with the aim of establishing strong new institutes which would unite geophysics and meteorology (Fig. 6.10). Soon afterwards, an interdisciplinary Institute for Geophysics, Geology and Meteorology (Institut für Geophysik, Geologie und Meteorologie) was created (unofficially, as the result of a basic initiative) under Franz Jacobs with the mission of ensuring better basic conditions during the renewal process at the university (Fig. 6.11).

Kollegium
zur Förderung der Geowissenschaften
an der Universität Leipzig

Leipzig, im Juli 1990

DENKSCHRIFT ZUM STAND UND ZUR ENTWICKLUNG

DER GEOWISSENSCHAFTEN AN DER UNIVERSITÄT LEIPZIG

Getragen von der Sorge um den Stand, die Entwicklung und die Wettbewerbsfähigkeit der Geowissenschaften an der Universität Leipzig erlaubt sich das am 25.6. 1990 gebildete Kollegium zur Förderung der Geowissenschaften an der Universität Leipzig, Ihnen diese Denkschrift mit der höflichen Bitte um Kenntnisnahme und Unterstützung zu überreichen.

Vor allem durch die Forscher- und Lehrerpersönlichkeiten wie den Mineralogen und Geologen C. F. Naumann, die Geologen H. Credner und F. Kossmat, dem Mineralogen, Petrographen und Kristallographen F. Rinne, den Geophysiker L. Weickmann, den Meteorologen V. Bjerknes, die Geographen von Richthofen und J. Partsch genossen die Geowissenschaften an der Universität Leipzig bis zum Ende des zweiten Weltkrieges einen bedeutenden internationalen Ruf. Jahrzehntelang bestand gleichzeitig eine unmittelbare Verbindung der Universität zu einer der wichtigsten Hoheitsaufgaben des Staates - zur geologischen Landesaufnahme, zur lagerstättenkundlichen und hydrogeologischen Ressourcenaufnahme mit dem Ergebnis, daß Sachsen ein halbes Jahrhundert lang zu den am besten erforschten Ländern der Erde zählte.

Als Folge des zweiten Weltkrieges, vor allem aber der dritten Hochschulreform 1968/69, verloren die Geowissenschaften an der Universität ihre einstige Bedeutung und ihre fünf Institute die strukturelle Selbständigkeit. Das Geographische Institut wurde nach Halle, das Geophysikalische Institut nach Berlin umgesetzt. Das Geologisch-Paläontologische Institut mit Museum und das Institut für Mineralogie und Petrographie wurden als selbständige Einrichtungen aufgelöst. Sie wurden mit dem Institut für Geophysikalische Erkundung einschließlich der Observatorien Collm/Sachsen und Zingst/Mecklenburg als Wissenschaftsbereich Geophysik der Sektion Physik bzw. als Wissenschaftsbereich Kristallographie der Sektion Chemie angegliedert.

Unser Anliegen besteht darin, ausgehend vom gegenwärtigen Bestand, die Geowissenschaften auf ein international beachtliches Niveau in Lehre und Forschung wieder anzuheben.

Um gegen andere deutsche und ausländische Hochschulen bestehen und den weitgefächerten geowissenschaftlichen Lehrbedarf an anderen naturwissenschaftlichen Einrichtungen befriedigen zu können, z. B. in den biologisch, agronomisch und ökologisch orientierten Wissenschaften, erscheint es notwendig, daß an der Universität Leipzig folgende Einrichtungen lehren und forschen:

Institut für Kristallographie, Mineralogie und Materialwissenschaften
Institut für Geologie und Paläontologie und Museum
Institut für Geophysik und Meteorologie
Institut für Geographie und Geoökologie.

Um den gestiegenen Anforderungen an die Geowissenschaften gerecht zu werden, ist auf längere Sicht ein Institutsverbund im Sinne einer geowissenschaftlichen Fakultät mit gemeinsam genutzten Einrichtungen zweckmäßig.

Die Neugründung und Modernisierung ist nicht nur aus der Sicht einer optimalen gegenseitigen Ergänzung in der Lehre der Geowissenschaften an der Landesuniversität Leipzig zu sehen. Im künftigen Land Sachsen wird ein hoher Bedarf an Geowissenschaftlern bestehen, die mit den speziellen geowissenschaftlichen und ökologischen Problemen eines industriellen und landwirtschaftlichen Ballungsraumes vertraut sind und zu ihrer Lösung beitragen können. Wir denken an Fragen der Erforschung und Sanierung des Lebensraumes, der Bergbaufolgelandschaft, der Trinkwasserversorgung und des Trinkwasserschutzes, der Reinhaltung der Luft, des sorgsamen Umgangs mit dem vielfältigen und komplexen Rohstoffpotential des Landes (Braunkohle, Steine und Erden, Wasser), der Entsorgung von Industrie und Kommunen einschließlich Recycling und Bewältigung der Altlasten.

Dr.habil.L.Eißmann, Geologe
(WB Geophysik)

Doz.Dr.sc.P.Schreiter, Mineraloge
(WB Kristallographie)

Dr.sc. F.Jacobs, Geophysiker
(WB Geophysik)

Dr.sc.W.Schmitz, Mineraloge
(WB Kristallographie)

Dr.A.Müller, Geologe
(Sächs.Akad.d.Wissenschaften)

Prof.Dr.sc.H.Neumeister, Geograph
(Akad.d.Wissenschaften der DDR)

Figure 6.10: "Kollegium zur Förderung der Erdwissenschaften". First Statement Juni 1990.

UNIVERSITÄT LEIPZIG
FACHBEREICH PHYSIK
WB/ **INSTITUT FÜR GEOPHYSIK, GEOLOGIE UND METEOROLOGIE**

IGGM Talstraße 35 D-O-7010 Leipzig

Deutsche Geophysikalische Gesellschaft
Herrn Dr. H. Jödicke
Correnssstraße 24
W - 4400 Münster

Telefon: (Leipzig) 71 65 0
Telefax: (Leipzig) 6858221

Ihr Zeichen Ihre Nachricht vom 21.2.92 Unser Zeichen Datum 6. März 1992

Betr.: Vorlesungsankündigung für Sommersemester 1992

Sehr geehrter Herr Jödicke,

im Sommersemester 1992 werden von uns folgende Lehrveranstaltungen angeboten:

Physik der Erde II	2V 1Ü	Jacobs
Allgemeine Geologie II	2V 1Ü	Fricke
Einführung in die Meteorologie	2V 1Ü	Wendisch
Meß- und Beobachtungstechnik der Meteorologie	1V 1Ü	Herber
Geophysikalisches Seminar	2S	Jacobs/Meyer

Mit freundlichen Grüßen

Figure 6.11: Institute for Geophysics, Geology and Meteorology. Education list 1991.

Backed up by the recommendations of the *Wissenschaftsrat der Bundesrepublik Deutschland* regarding "The development of the natural sciences in the New States of the Federal Republic of Germany" and on the basis of the *Sächsisches Hochschulerneuerungsgesetz* (Saxon University Renewal Law) of August 4, 1993, the University of Leipzig committed itself to re-establishing teaching in the earth sciences in their entirety, and to pursuing highly-qualified research in selected areas.

On June 1, 1992 the Chair of Geophysics was awarded to Franz Jacobs as one of the key professorships in the context of an extraordinary appointment procedure undertaken by the *State Ministry for Science and Art (Ministerium für Wissenschaft und Kunst)* of the Free State of Saxony. At the same time, the ministry set the formal requirements for filling all the professorships in geophysics,

Figure 6.12: 52. Annual Conference of the German Geophysical Society (DGG), Leipzig 1992. Program (cover).

meteorology, geology and geography with associated staff appointments plans at the University of Leipzig and also non-university professorships.

In March 1992, the 52nd Annual Conference of the German Geophysical Society (DGG) was held in Leipzig as an expression of the new possibilities for joint work and research in a reunified Germany (Figures 6.12, 6.13).

In the history of the Department of Geophysics within the Section of Physics, it will remain in memory that, apart from the progress made in research, the staff, driven by "a community of fate", found a way to make co-operation between geophysicists and meteorologists productive and successful. This was particularly true in finding uses for the new possibilities opened up by digital data recording and processing, and in a further sense also in organizing the new structures at a time of great political change.

Figure 6.13: 52. Annual Conference of the German Geophysical Society (DGG), Leipzig 1992. Opening session (from left, first row: H. A. K. Edelmann, R. Lauterbach, W. Zettel; second row: J. Wohlenberg, R. Hänel, J. Gruntorad.

The Institute of Geophysics and Geology since 1993

Franz Jacobs and Michael Korn

December 2, 1993 was the 584th anniversary of the foundation of the University of Leipzig "Alma mater Lipsiensis" and also the day several new faculties and institutes were established. They originated out of the Faculty of Mathematics and Sciences that had been revived (after 1990) under the Deanship of Franz Jacobs. Among the new faculties and institutes was the Faculty of Physics and Earth Sciences and, within it, the Institute of Geophysics and Geology (IGG) / the Geological-Palaeontological Collection (Fig. 7.1). The institute hosted the following geo-

UNIVERSITÄT LEIPZIG

Auf der Grundlage des Sächsischen Hochschulgesetzes
vom 4. August 1993

wird

unter dem Rektorat des Professors für Theoretische Chemie
Dr. rer. nat. habil. Cornelius Weiss

das

Institut für Geophysik und Geologie

gegründet.

Leipzig, am 2. Dezember 1993

C. Weiss

Der Rektor

Figure 7.1: The Institute of Geophysics and Geology. Establishment document. 1993.

Figure 7.2: Helmut Meyer (1947–2011).

physics professorships: Physics of the Earth (Franz Jacobs, first director), Theoretical Geophysics (Michael Korn), and Engineering and Environmental Geophysics (Helmut Meyer; Fig. 7.2). Since 1993, the Collm Geophysical Observatory has been shared with the Institute for Meteorology.

During the 1990s, geophysics developed rapidly thanks to considerable support from the state and as a result of the high levels of motivation among professors, scientific and technical staff and students. Enrolments in geophysics and geology rose to about 70–80 per year, success in attracting external funding allowed dozens of PhD students to be employed. At the same time, the technical apparatus was upgraded and the less than satisfactory state of repair of the buildings at Talstraße 35 and the Collm Geophysical Observatory were considerably improved by means of third-party funding and financial support from the Free State of Saxony.

Several main research themes were established within international networks:

- Geophysics at deep drilling sites for investigation and usage of geothermal energy, earthquakes und palaeo-climate (KTB continental deep drilling site, geothermal and research drill holes in Gross-Schönebeck and Baruth) (Fig. 7.3)
- Geoelectrical tomography with projects in abandoned mining areas, in the Long Valley Caldera (California) and at Mount Merapi (Figures 7.4, 7.5, 7.6)
- Seismology and seismic monitoring (Saxonian Seismological Network)
- Geophysics at geological barriers (subsurface waste deposits, final storage sites in salt domes, flooding embankments in Saxony) (Fig. 7.7)
- Seismic wave propagation in complex structures
- Volcano seismology
- Seismic tomography on different scales

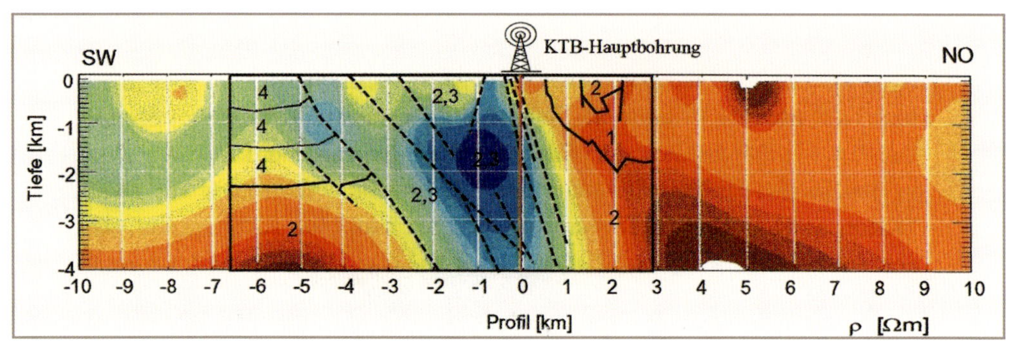

Figure 7.3: German Continental Deep Drilling Project (KTB). Electrical Resistivity Tomography (ERT), 1994.

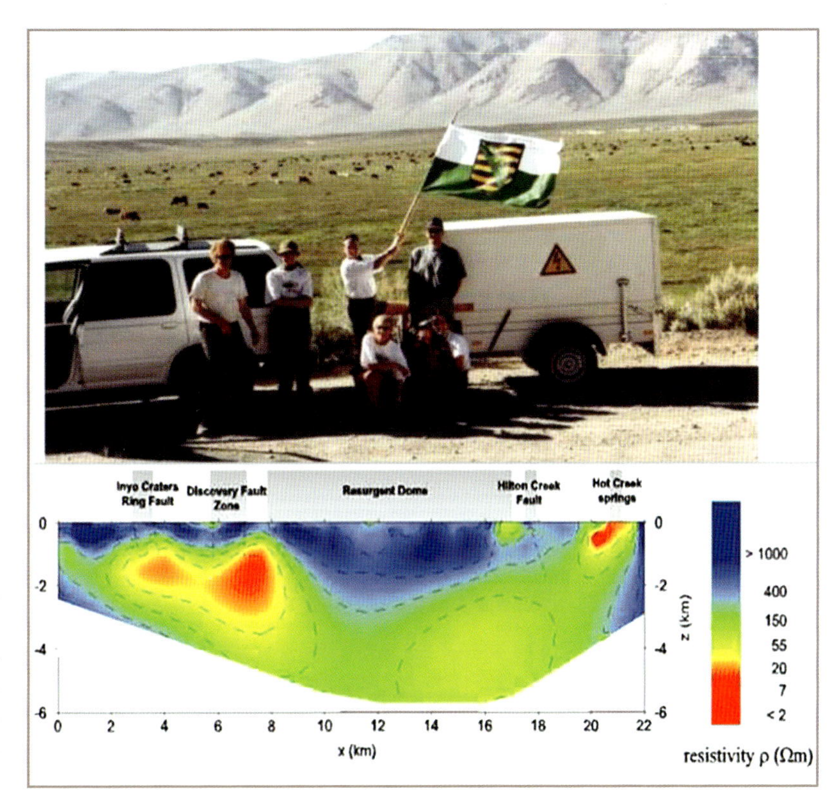

Figure 7.4: Long Valley Caldera (California). The team and the geophysical resistivity distribution in the underground. 2000.

Figure 7.5: Geophysical monitoring station. Summit Volcano Merapi (Indonesia), 2900 m asl, 2001.

Figure 7.6: Volcano Merapi. Finite-Element Model and Electrical Resistivity Tomography (ERT).

Hochwasserschutz durch geoelektrische Deichdiagnose

Protection of floods by geoelectrical dike diagnostics

Multielektrodensystem auf einem Elbedeich bei Torgau ...

Hält der Deich?

- Bauweise
- Material
- Durchfeuchtung

sind mit elektrischer Leitfähigkeit verknüpft

Zerstörungsfreie Diagnose?

- Geoelektrik
- Multielektrodensysteme
- Leitfähigkeitstomographie

Technische Ausrüstung?

- Messapparatur, tragbar, batteriebetrieben, programmgesteuert
- Edelstahlelektroden, vieladrige Kabelbäume
- Interpretationssoftware

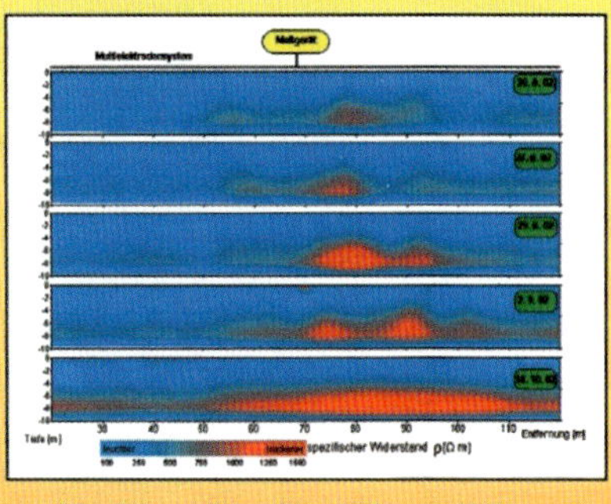

... und Ergebnisse nach dem Augusthochwasser 2002

Figure 7.7: Protection of floods by geoelectrical dike diagnostics, 2002.

Of special interest were the annual workshops on high resolution geoelectrics that took place near Leipzig since 1992. They became known as the Bucha Seminars, and attracted about 50 participants each year who discussed the progress of electrical resistivity methods in science and practical application (Fig. 7.8).

Since 1996, the institute published *Leipziger Geowissenschaften*, a journal appearing in 1–2 volumes per year (Fig. 7.9) in continuation of the tradition of publishing geophysical-geological articles written by researchers at the University of Leipzig. The journal presents pieces on geophysics, geochemistry, geology, palaeontology, geography, environmental science and history of earth sciences and has been edited by, among others, Arnold Müller, Michael Börngen and Frank Bach.

In spite of recognized achievements, explicit support from the City of Leipzig and initially unequivocal public affirmations from the university

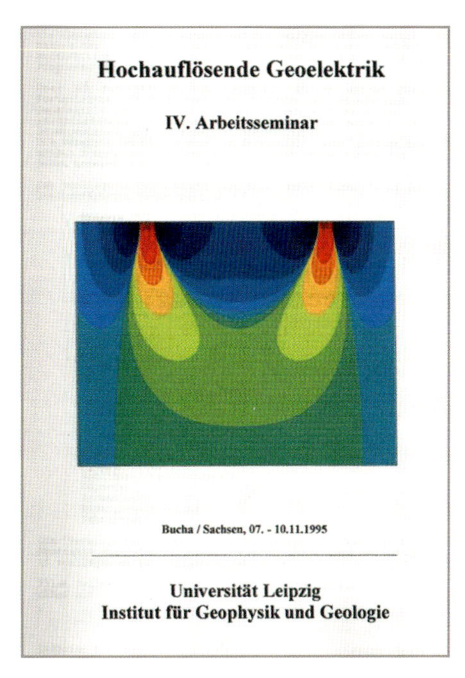

Figure 7.8: Bucha Seminar on high resolution geoelectrics. Program (Cover) 1995.

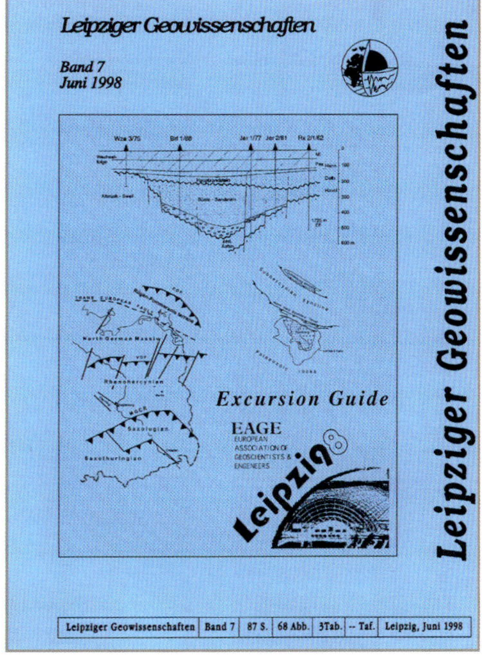

Figure 7.9: Journal „Leipziger Geowissenschaften". Volume 1, 1996 and Volume 7, 1998. Covers.

Table 7.1: Number of first-semester students for diploma and MSc study programmes since 1990.

	Diploma Geophysics	Diploma Geology	Msc Geosciences
1990	7		
1991	2		
1992	2		
1993	6		
1994	15		
1995	14	9	
1996	11	24	
1997	9	18	
1998	8	35	
1999	14	32	
2000	6	21	
2001	9	45	
2002	14	68	
2003	16	75	
2004			
2005			
2006			
2007			
2008			2
2009			10
2010			10
2011			11
2012			18

and the faculty, 2004 brought deep cuts in budget and personnel that were necessitated by deteriorating financial circumstances. The study programmes in geophysics and geology were closed, staff numbers were reduced by more than 50 per cent and two of the three geophysics professorships were eliminated. As a consequence, nationally and internationally recognized successful and socially relevant research profiles closed down. In 2004, Franz Jacobs retired, and the new director of institute became the geologist Werner Ehrmann. Theoretical geophysicics specializing in seismology and including the Collm Observatory remained the only stable research area in geophysics.

In 2007/2008, a new MSc course entitled „Geowissenschaften: Umweltdynamik und Georisiken" was established. Specialising in environmental dynamics and geohazards, the programme builds on the long tradition and expertise in geological and geophysical research and teaching in Leipzig. The reconstruction of environmental dynamics on geological time scales and the investigation of geohazards are topical themes in earth science. As a non-consecutive Msc course this programme is open to students with various backgrounds in geosciences coming from different universities. The annual number of first-semester students and of graduations in the various study programmes are listed in Table 7.1, and 7.2, respectively.

At present, geophysics focuses on risks and opportunities of using subsurface space in exploiting energy and natural resources, the storage and deposit of dangerous wastes, and monitoring subsurface dynamic processes. This includes earthquake monitoring and hazard assessment in central Germany, the use of geophysical techniques for discovering, securing and the long-term monitoring of underground waste deposits (Fig. 7.10), the safe development and usage of geothermal energy, and carbon dioxide capture and storage. The expertise available at the University of Leipzig in methodological developments as well as the broad technical and instrumental capabilities (seismics, radar, geoelectrics) allows a wide range of projects in both applied and basic sciences to be carried out.

Among ongoing research activities are seismological and geoelectrical investigations of fluid migration in the West Bohemia/Vogtland earthquake swarm area within the framework of the International Continental Scientific Drilling Program (ICDP), the establishment of a permanent earthquake warning service for Saxony (Fig. 7.11), and the development of seismic interferometry for imaging and real-time monitoring for the a variety of applications on different scale-lengths ranging from material testing and geotechnical applications to monitoring of volcanoes and active tectonic areas.

The Institute of Geophysics and Geology has numerous co-operations with national and international research institutions, universities and authorities. Among them are: the Helmholtz Centre for Environmental Research (UFZ); Helmholtz Centre Potsdam, German Research Centre for Geosciences (GFZ); the Alfred Wegener Institute for Polar and Marine Research (AWI); Helmholtz Centre Dresden-Rossendorf (HZDR); Leibniz Institute for Applied Geophysics (LIAG); Universidad de Concepcion, Chile; Tohoku University, Japan; Czech Academy of Sciences, Charles University Prague.

Table 7.2: Number of diploma, MSc, doctoral, and habilitation graduations since 1990 (only Geophysics).

	Diploma	MSc	Doctorate	Habilitation
1990	7		4	1
1991	4		3	1
1992	5		5	
1993			1	
1994	1			
1995	4			
1996	7			
1997	3		1	
1998	2		3	
1999	8		3	1
2000	7		1	
2001	2		2	
2002	5		1	
2003	2		3	
2004	11			
2005	3			
2006	2			
2007	2		2	
2008	3			
2009	5			
2010			3	
2011	2	2		
2012		3		

Figure 7.10: Project Salinargeophysik. Measurements in a salt mine for investigation of geological barriers, 2007.

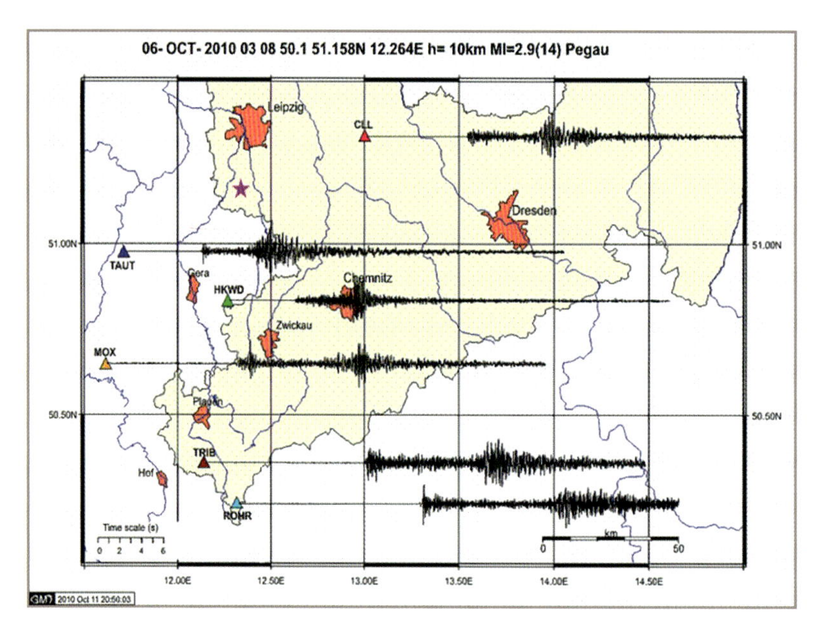

Figure 7.11: Saxony Seismological Network. Seismogram displaying a seismic event near Pegau (ML=2,9), 2010.

The Institute for Meteorology since 1993

Manfred Wendisch and Christoph Jacobi

On the occasion of the 100th anniversary of the foundation of the Geophysical Institute at the University of Leipzig, it seems appropriate to summarize the history and major achievements of the Institute for Meteorology (the so-called Leipzig Institute of Meteorology, LIM) over the last two decades. LIM was officially established in 1993 (Fig. 8.1). The first director was G. Tetzlaff, who accepted a call for a professorship at the University of Leipzig in 1993. At the time, the re-organization of the University of Leipzig following the reunification of Germany was still incomplete; it included changes in the management structures of the university, which meant a shift from the former structure in sections (Sektionen), via an intermediate stage of divisions (Fachbereiche), to the current structure of faculties and subordinate institutes.

Two factors helped introduce and establish LIM as a separate meteorological institute at the University of Leipzig. First, a long-standing tradition of meteorological teaching and research did already exist at the University of Leipzig, starting with the prominent meteorologist and atmospheric physicist (explorer of the polar front theory) Vilhelm Bjerknes, who held the founding chair of the Geophysical Institute 1913–1917 (see Chapter 1). The Geophysical Institute built on an even longer tradition reaching as far back as Virgil Wellendarffer's "Decalogium" in 1507, one of the very first meteorological textbooks. Second, there was political support for rebuilding the meteorological teaching and research infrastructure in the eastern part of Germany after the German unification. This political intention was also the basis for the foundation of the Leibniz Institute for Tropospheric Research (Leibniz-Institut für Troposphärenforschung, IfT,

UNIVERSITÄT LEIPZIG

Auf der Grundlage des Sächsischen Hochschulgesetzes
vom 4. August 1993

wird

unter dem Rektorat des Professors für Theoretische Chemie
Dr. rer. nat. habil. Cornelius Weiss

das

Institut für Meteorologie

gegründet.

Leipzig, am 2. Dezember 1993

C. Weiss
Der Rektor

Figure 8.1: Official document establishing the Institute for Meteorology.

Figure 8.2: Main building of the Institute for Meteorology (LIM) at Stephanstraße 3 in Leipzig.

now called TROPOS), which was established in 1992. The idea was that IfT, as a generously-funded, non-university institute, and LIM, as an institute within the university, would closely collaborate and support each other. The focus of LIM would be teaching students in undergraduate and graduate meteorological courses, whereas IfT would concentrate on research involving the students trained at LIM. Also the university would provide the forum for doctorates and habilitations. The faculty in which LIM was embedded is responsible for providing the basis necessary for the university part of the meteorology study programme.

Lectures in meteorology started in 1990 when nine students enrolled (Table 8.1). At that time, the responsibility for teaching rested with W. von Hoyningen-Huene and the research assistant A. Raabe. M. Wendisch and A. Herber (both PhD students at that time) were also involved in the teaching. In 1991 and 1992, student enrolments dropped to just five each year, but the existing programme and the students nonetheless demanded that the necessary institute structures were assembled. Only if the personnel and material capacities were provided by the university could the study programme in meteorology be legally secured on a long-term basis.

From the beginning, it was obvious that LIM needed to be established as an independent organizational unit in close proximity to the faculty's physics institutes. A concept for the personnel for the new institute was put together, based on existing resources. These resources were distributed between facilities in Leipzig, the Collm Geophysical Observatory about 60 km east of Leipzig (head scientist R. Schminder), and the Zingst Maritime Observatory on the Baltic coast (headed by H.-J. Schönfeldt). At that time, the total staff (including the external facilities at Collm and Zingst) comprised one professor, five senior scientists, eleven technicians, and one full time and one part-time administrator. However, the re-structuring at the university combined with a need to balance scientific and technical staff meant that five of the technician positions were slated to be cut within a couple of years. Of the remaining 12.5 positions, less than half were based in Leipzig, and only two were involved in teaching in the meteorology programme.

G. Tetzlaff started lecturing in October 1992. By the end of the year, the university authorities had approved bringing the personnel in Leipzig under one roof, and rooms were made available in two buildings at Stephanstraße 3 (Fig. 8.2). An administrative unit for managing LIM was established; the responsibility for the facilities in Zingst and for the upper atmosphere department at the Collm Geophysical Observatory were transferred to LIM. On the 1st of March in 1993, G. Tetzlaff was officially appointed Professor of Meteorology at the University of Leipzig.

Table 8.1: Number of first-semester students for diploma study program in meteorology, bachelor study program "*" respectively.

Year	Students
1990	9
1991	2
1992	3
1993	10
1994	14
1995	14
1996	26
1997	29
1998	36
1999	40
2000	41
2001	43
2002	45
2003	68
2004	97
2005	109
2006*	31
2007*	98
2008*	71
2009*	67
2010*	64
2011*	46
2012*	70

In spring of 1993, there were ten undergraduate students in meteorology, including a few in the first year (Table 8.1). The most urgent challenge was to procure lecture rooms and the equipment for the laboratory and field courses needed in the student's practical training. Furthermore, examination regulations had to be drafted, discussed and approved by the faculty and the university administration. The regulations were drawn up based on experience at other universities. This turned out to be a relatively straightforward process and the results were quickly approved by all the relevant authorities.

Another important step in 1993 was to physically move LIM into the buildings at Stephanstraße 3. One of the two buildings had previously been utilized as electrical and mechanical workshops, though without specific organizational attribution. Half of the ground floor had been converted to residential use, and it took a couple of years before the tenants could be convinced to move out. The workshop fittings, including heavy mechanical machinery, were removed, and the chancellor of the university agreed to renovate the buildings. As a complete refurbishment program for the entire building would take a long time to plan and execute, it was decided to proceed stepwise. This took longer than expected and the process was finished only after several years. In the second building at Stephanstraße 3, the workshop staff started to be withdrawn gradually, until eventually the entire building was available for LIM.

In summer 1993, the meteorological research units in Zingst and Collm were officially declared dependent external sites as part of LIM, and the personnel were relocated to Leipzig. In the case of Collm, the continuation of the unique series of upper atmosphere wind observations was safeguarded initially by automating the measurement process and installing new computer technology, later by replacing the existing measurement systems with state-of–the art wind radar which could operate largely unattended. Particular challenges in Zingst included the upkeep of oceanographic and coastal meteorology research equipment, in particular a boat (and crew), a boat house, and a mooring dock. The resources needed to maintain these facilities far exceeded the overall LIM budget, with the result that continuing experimental work in Zingst would require third-party project funding. Numerous applications were submitted, but the resources forthcoming nevertheless did not provide sufficient support for the infrastructure in Zingst over the long term, and so it had to be abandoned. Nowadays, the facilities in Zingst still belong to the University of Leipzig (though not as part of LIM) and are used for advanced practical courses for bachelor students of meteorology, among other things.

On 2 December 1993, LIM was officially founded (Fig. 8.1), and was incorporated into the Faculty of Physics and Earth Sciences. The study programme in meteorology was finally approved, and it was forecast that the number of students would grow in the future. Indeed, by the 2000s the number of freshers had increased to more than 50.

Very soon it became clear that, even with the relocation of the personnel from Zingst and Collm to Leipzig, the teaching capacity would not be sufficient. It was necessary to establish a professorship for Theoretical Meteorology. The new professorship was advertised before the end of the year and the first holder, W. Metz, took up his post at the beginning of the summer semester of 1994. Research assistant and later Junior Professorship positions were held by U. Harlander, T. Trautmann, R. Faulwetter, A. Ziemann, and B. Pospichal. Maintaining these positions was quite an achievement, both scientifically and with regard to securing teaching capacity. In 1999, R. Schminder, who represented (together with D. Kürschner) upper atmosphere research at Collm, retired. In order to save this branch of LIM's research and the unique series of observations on upper atmosphere winds, a new professor's position was created and filled initially by Ch. Jacobi, who had joined the upper atmosphere working group in 1995 and successfully finished his habilitation in 2000. Just a year later, Ch. Jacobi was officially appointed to the new professorship, becoming the third professor at LIM in 2001.

Besides developing the teaching capacities at LIM, the new meteorological institute had to establish a solid position for itself within the meteorological community – nationally, internationally, and within the university. It proved an advantage that the members of the institute had different scientific roots, experimental and theoretical, ranging from the lower to the upper atmosphere. Scientific areas covered were climate change statistics, boundary layer meteorology and regional climatology including mesoscale modelling, wind energy research, theoretical wave dynamics, remote sensing, coastal processes, upper atmosphere dynamics, and atmospheric acoustics. The journal "Wissenschaftliche Mitteilungen aus dem Institut für Meteorologie" was established in 1995 as a publication forum for these different research directions (Fig. 8.3). This series has been edited by A. Raabe for most of the intervening years until today. The journal appeared several times a year, and functioned as a platform for the publication of doctoral and habilitation theses. Later, the scientific fields covered by LIM were extended to include clouds and radiation, climate dynamics, and remote sensing.

Figure 8.3: Cover of the "Wissenschaftliche Mitteilungen aus dem Institut für Meteorologie".

LIM established national and international scientific collaborations. The most close collaboration evolved with IfT (now TROPOS). The director of TROPOS and the two heads of departments are official members of LIM and the faculty and are appointed professors at the University of Leipzig. TROPOS professors and senior scientists significantly contribute to teaching at LIM, bringing their scientific competence and experience to the students. Contacts with Umweltforschungszentrum (UFZ) Halle-Leipzig developed, based on shared scientific interests in the study of urban environments. The contact to Forschungszentrum Rossendorf grew out of a mutual interest in wind energy use. These contacts have made it possible to establish a series of specialised lectures held by personnel from these local organizations. Another important contact has been established with the German Weather Service (Deutscher Wetterdienst, DWD), which includes lectures on and practical experience with synoptic meteorology in the undergraduate and graduate study programmes. Also, numerous research cooperations and projects have evolved, resulting in a number of joint publications (Table 8.2). Several projects of an applied nature have been undertaken with

Table 8.2: Number of funded scientific projects active in any one year, and number of peer reviewed papers (source: annual research report to the University of Leipzig).

	Number of projects	Number of papers
1993/4	15	22
1995	16	23
1996	20	39
1997	21	42
1998	13	39
1999	17	36
2000	17	36
2001	17	30
2002	20	37
2003	23	40
2004	21	44
2005	16	34
2006	19	44
2007	17	29
2008	17	25
2009	22	40
2010	23	35
2011	23	40

local industry and public institutions in Saxony, while collaborations, mainly on basic research topics, have grown up with many other national and international meteorological university institutes and extra-university research bodies. These include, for example, DLR (Deutsches Zentrum für Luft- und Raumfahrt), GFZ (Geoforschungszentrum) Potsdam, AWI (Alfred Wegener Institute for Polar and Marine Research) Bremerhaven, Max-Planck institutes in Mainz, Jena, and Hamburg, Fraunhofer institutes, and other partners all over the world (NASA and NOAA in the USA, Meteo France, the MetOffice in the UK, and numerous foreign universities).

The number of graduates in meteorology from LIM grew year by year (Table 8.3). The first diploma degrees were awarded in 1995, and from then on the number of degrees awarded (Master's after 2006) rose gradually, reaching 25 in 2010. Table 8.3 also shows the number of doctorates awarded. The majority of these came from the collaboration with TROPOS. The total number of diploma/Master's degrees in the years between 1995 to 2011 was 157; the total number of doctorates in the same period was 62. Altogether, 7 habilitations (postdoc qualifications) were completed at LIM (Table 8.4).

The year 1999 saw the launch of the so-called Bologna Process. Among other changes, it gave students the option of concluding their studies after three years with a Bachelor's degree; however, the vast majority of students opt to continue their studies within the Master's program after completing the Bachelor's degree. When the requirements of the two-step programme for Bachelor and Master of Sciences degrees were taken together, it transpired that ultimately the previous diploma comprised a very similar curriculum, meaning that the diploma degree was in fact very similar to the new Master's degree. As far as the students and the quality of their degrees are concerned, the effect is mainly that the scope of the examined subjects is more clearly arranged and the content of modules is more clearly described.

In 2008, G. Tetzlaff retired from the professorship in meteorology. In the following two years, five (out of seven) professorships at LIM and TROPOS were re-appointed. Tetzlaff was succeeded in 2009 by M. Wendisch, who specialised

Table 8.3: Number of diploma, doctoral, and habilitation graduations per year in meteorology.

	Diploma	Doctorate	Habilitation	BSc	MSc
1995	4	2	–		
1996	3	2	–		
1997	1	3	–		
1998	4	7	–		
1999	5	2	2		
2000	9	8	1		
2001	3	3	–		
2002	6	5	1		
2003	5	4	2		
2004	11	4	–		
2005	12	4	–		
2006	12	4	1		
2007	14	3	–		
2008	14	2	–		
2009	18	3	–	9	–
2010	25	2	–	20	–
2011	11	5	–	19	2

Table 8.4: Habilitations

1999	A. Ansmann, N. Mölders
2000	C. Jacobi
2002	U. Harlander
2003	U. Schlink, M. Wendisch
2006	T. Lawrence

in atmospheric radiation, a scientific field already covered by T. Trautmann (Fig. 8.4). B. Pospichal joined LIM as the new Junior Professor for ground-based remote sensing in 2010. The professorship for Theoretical Meteorology held by W. Metz, who retired in 2010, was upgraded (from C3 level to W3) and filled in 2011 by J. Quaas, who is actively involved in global climate modelling. The upgraded professorship enhances the teaching at LIM and adds to research activities at TROPOS. After the retirement of the first Director of IfT, J. Heintzenberg, A. Macke was appointed new Director of TROPOS in 2010. Furthermore, I. Tegen replaced E. Renner as the new head of the department for atmospheric modelling at TROPOS in 2012. With the arrival of these new professors, meteorological teaching and research in Leipzig will undoubtedly take on a new dynamic in the coming years.

New research areas covered within Leipzig meteorological research are mostly centred on the role of clouds (ranging from low-level boundary layer and mixed-phase clouds to upper tropospheric cirrus) within the Earth's climate system

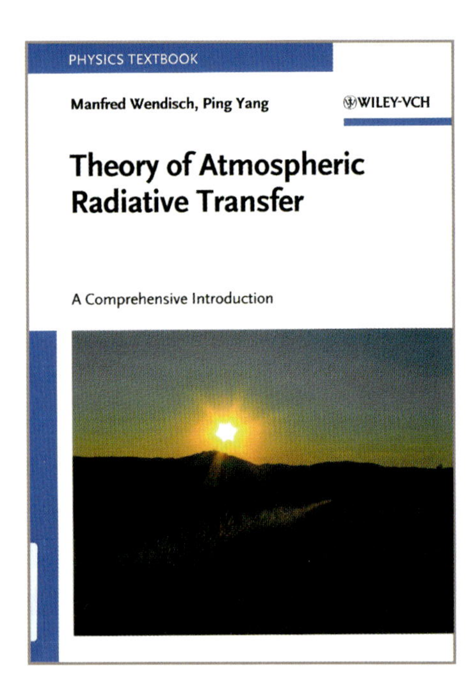

Figure 8.4: Book on atmospheric radiation published by Wendisch and Yang in 2012.

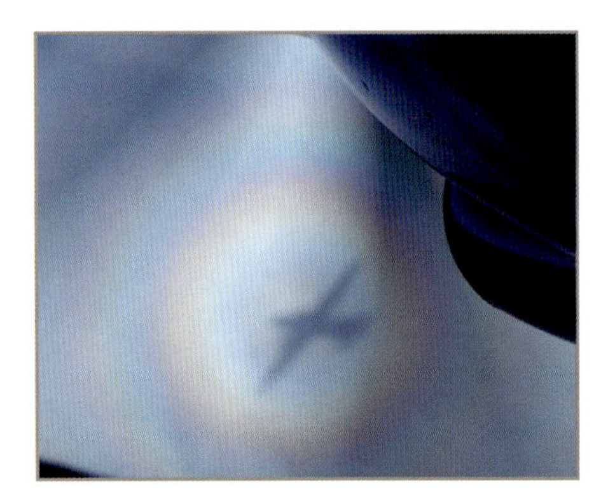

Figure 8.5: Aircraft measurements to study the effects of mixed-phase clouds in the Arctic using the Polar-5 research aircraft operated by AWI. Glories like the one shown in the photo are products of light being scattered by water droplets. Their structure contains information about droplet size distribution within the top layer of the cloud.

(Fig. 8.5). Their importance for radiative and dynamic processes in the boundary layer, the troposphere, up to the high atmosphere, from the local to the global scale, will be one of the foci of future meteorological research at LIM and TROPOS. These processes will be studied using experimental and modelling methods. The experimental facilities available include laboratory and field observations (ground-based and airborne, in situ and remote sensing). The simulation capacities available in Leipzig at LIM and TROPOS include models from high resolution to global scales. Furthermore, aerosol processes (formation, evolution, impact on cloud life time) will be studied in close collaboration between LIM and TROPOS in Leipzig. Recently, a new graduate school to study the effects of desert dust in the atmosphere was initiated by TROPOS; it involves LIM and further university partners. Within an ERC (European Research Council) starting grant awarded to J. Quaas (LIM) the radiative forcing by aerosol-cloud effects is investigated on a global scale. Furthermore, currently LIM coordinates the DFG (Deutsche Forschungsgemeinschaft) priority program SPP 1294 (Schwerpunktprogram) Atmospheric and Earth System Research with HALO (High Altitude and Long Range Research Aircraft). These exciting new research topics will contribute to maintaining and strengthening the position of Leipzig meteorological research in the international community, and will continue to attract students and young researchers in the future.

The Collm Geophysical Observatory

Michael Korn and Christoph Jacobi

Ludwig Weickmann, who became head of the Geophysical Institute in 1923, wanted to build a geophysical observatory outside Leipzig that would be dedicated primarily to seismology, because the conditions for seismological observations within the city had greatly deteriorated as a result of increasing industrialization. In 1927, the Saxon Ministry of Education approved the construction of an observatory that, despite the difficult economic conditions at the time, was then erected on the slopes of Collm hill, 6 km west of the town of Oschatz, and about 60 km east of Leipzig. The groundbreaking ceremony was held on May 9, 1931, and the topping out ceremony on August 5, 1931

Figure 9.1: Topping-out ceremony of the main building at Collm Geophysical Observatory, August 5, 1931 with Prof. Weickmann at top left and Prof. Moltschanoff at bottom left.

(Fig. 9.1). The purpose of the observatory and the reasons for choosing this particular site are described in the founding document that was embedded in the observatory tower:

For the planning of the observatory, it was crucial to achieve favourable working conditions for special branches of geophysical science, such as seismology, magnetism, gravity measurement, atmospheric electricity investigation, radiation measurement and meteorological investigation. The location of the observatory on Collm hill near Oschatz, far from any disturbance by vibrations from railways, cars and mining blasts, free of influence by DC current on the electromagnetic fields, far from sources of atmospheric haze caused by smoke of industrial plants, will guarantee the possibility of perfect observations.

Besides the main building with working and living spaces, a mechanical workshop and a lecture hall, the observatory also had several additional buildings which accommodated the observational instruments (seismic, absolute gravity and magnetic huts). The official opening ceremony took place on October 9, 1932 on the occasion of the 10th annual meeting of the German Geophysical Society, and the observational buildings were erected and went into operation

Figure 9.2: Aerial photograph of the observatory (Postcard, E. Assmus, Leipzig, around 1938).

Figure 9.3: Old Seismometer hut ("Credner-Weickmann-Erdbebenwarte"; photograph taken 2012).

Figure 9.4: Document honouring the supporters of the Observatory.

by September 1934 (Figures 9.2, 9.3). The completion of the observatory was the final element in realising the envisaged extension of the institute's range of activities with regard to solid earth observations. A document dedicated to the supporters of the observatory is depicted in Figure 9.4.

Among the scientists who had a major impact on the research work and developments at the observatory were:

Paul Mildner
Hildegard Nitsche (by marriage Kohlsche)
Alfred Adlung
Hans Koch (1955–1969)
Christof Junge (1958–1965)
Rudolf Schminder (1959–1995)
Bernd Tittel (1965–2003)
Dierk Kürschner (1966–2006)
Siegfried Wendt (from 1978).

Starting in 1957, publication of the "Geophysikalische Meßreihen des Geophysikalischen Obervatoriums Collm und des Maritimen Observatoriums Zingst"

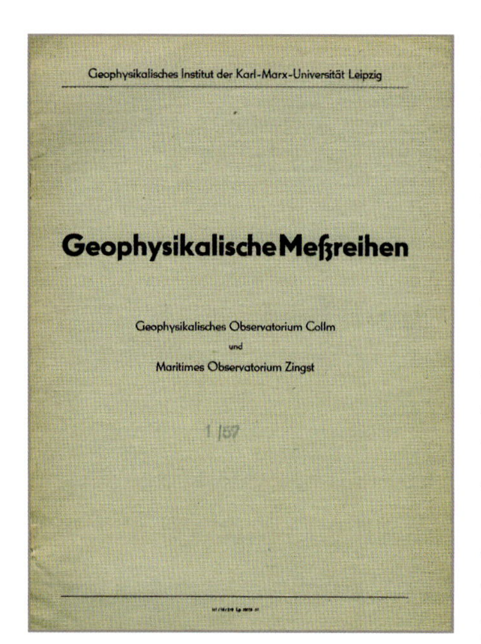

Figure 9.5: Cover of "Geophysikalische Meßreihen".

(Geophysical Measurements of the Collm Geophysical Observatory and the Zingst Maritime Observatory) began (Fig. 9.5), containing data series from magnetics, upper atmosphere research and seismology, among other information. The last printed publications appeared in 1980, with the monthly reports being published electronically since then.

Up to 16 people worked in the observatory during its peak, but today the number has decreased to three full-time or part-time jobs, due to progress in automatic acquisition instruments and data transfer.

The main building has been refurbished in recent years and is now in a very good state of repair. However, while seismic and upper atmosphere measurements are carried out continuously in the grounds of the observatory, the future use of the main building itself is under discussion.

Upper Atmosphere Research

In 1956, research on the physics of the upper atmosphere started at Collm with the goal of contributing to the International Geophysical Year. First projects were devoted to the measurement of lower ionospheric absorption, fixed frequency ionospheric measurements on 27 MHz, and investigation on the propagation of radio waves. However, soon the focus was placed on upper atmosphere wind measurement, which was advanced mainly by Rudolf Schminder, with routine data collection after 1959.

The wind measurement principle was based on determining the time differences between corresponding fading observed in sky wave registrations of low frequency (about 200 kHz) radio waves measured at closely spaced receivers situated on the corners of a right-angled triangle of 300 m side length. Since fading is caused by irregularities in lower ionospheric electron density which move with the prevailing wind, the measured time differences can be converted into wind vectors. Since the method was based on data taken from several receivers a small distance apart, it was called the "closely-spaced receiver method", while the use of low-frequency radio waves is documented by the annex "LF".

Given a distance between receivers of 300 m and a typical wind velocity of 10 m/s, the time differences between corresponding fadings amount to several

Figure 9.6: LF wind plotting system.

seconds. Because the applied mathematics in itself is simple, it was possible to apply manual data analysis during the first few years of wind measurements. Nevertheless, the manpower required was high, so that Dierk Kürschner tried in the following years to automate the analysis (Fig. 9.6), with the result that it was finally made fully automatic in 1972; while a sufficiently long parallel analysis was performed beforehand in order to avoid inhomogeneity as far as possible, as this would affect long-term analyses of the time series. The new analysis system allowed a significant extension of the daily measurements, limited only by strong absorption in the lowermost ionosphere (the "D-region") during daylight hours.

The measurements were carried out until 2007 and allow the analysis of long-term trends in connection with climate variability estimation.

From other measurements, it was known that the reflection of LF waves occurs at about 90 km at night, and at 80 km during daytime. However, in the long run, not knowing the reference height having to assume approximate and mean heights proved unsatisfactory. Therefore, the next step was to determine the reflection height, and this process was started in 1982 by means of phase comparison between the sky wave and the ground wave from the radio trans-

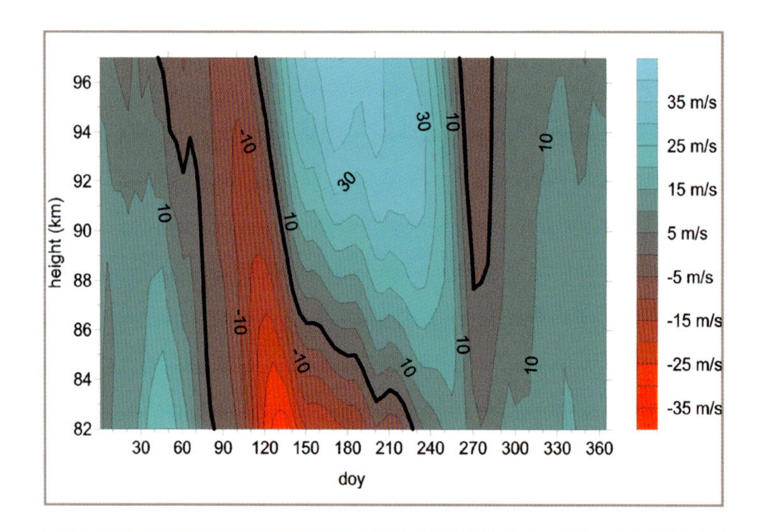

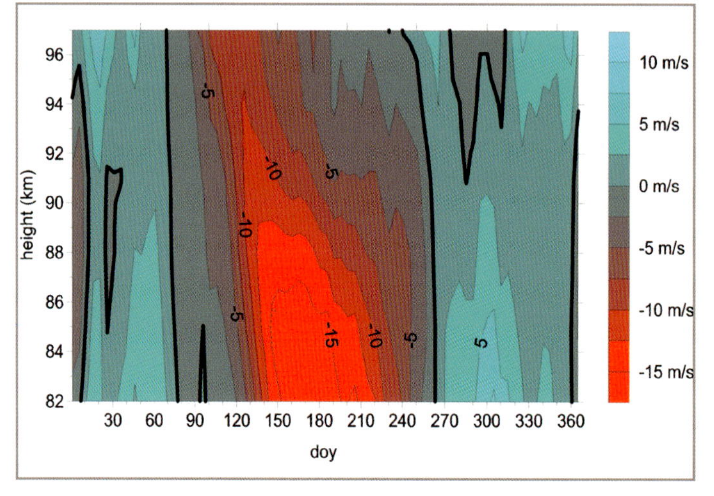

Figure 9.7:
Zonal (upper panel, positive eastward) and meridional (lower panel, positive northward) wind component according to meteor radar measurements 2004–2011.

mitters. These measurements made it possible to estimate monthly mean profiles of upper atmosphere wind in the height range between 80 and 95 km.

The LF measurement principle was only rarely used worldwide, because of the measuring gap during daytime and the fact that so-called meteor radars operating in the VHF frequency range deliver more reliable data at the same height range. Consequently, a VHF meteor radar was installed at Collm in 2004 to extend wind measurements. Figure 9.7 shows height-time cross-sections of the zonal (eastward directed) and meridional (westward directed) mean wind calculated from measurements in the period 2004–2011. In this figure, winter westerlies and summer easterlies are visible in the lower layers, as well as a transition to westerlies in summer at about 90 km altitude. The transition is connected with equator-ward meridional winds, obvious in the lower panel of Figure 9.7.

The two methods have been applied in parallel for 4 years. This was necessary because it turned out that LF measurements systematically underestimated wind amplitudes. On the basis of a comparison between the two methods, however, it is now possible to apply a correction to the LF registrations and to continue to use the Collm wind registrations for long-term analyses of upper atmosphere dynamics.

The Collm wind measurements are part of the work done by the upper atmosphere working group headed by Ch. Jacobi since 2001. Research topics include the analysis of global satellite data, the numerical simulation of global middle atmosphere circulation, ionospheric research and EUV measurements. However, the Collm wind measurements are still an important part of the work and will continue in the future.

Solid Earth Research

With the beginning of operations at the observatory, the instruments for measuring geophysical data began to be installed in the auxiliary buildings on site. The year 1935 saw the start of recording the variations in the Earth's magnetic field, and this continued for many decades with only few interruptions during World War II. However, the data never gained widespread significance, and the data collection finally ceased in 2000.

Instead, the observatory became much more important as the location for seismological data acquisition. The Wiechert seismograph was brought from Leipzig to the observatory in 1935, and it is still in operation today. Over the following decades, the observatory became home to several generations of seismic instruments, among them a set of short-period Benioff seismometers, two restored Wood-Anderson torsion seismographs, long-period VSJ-1 and short-period SSJ-2 seismo-

Figure 9.8: Top: Long period electrodynamic seismometer (VSJ-1, Jena). Bottom: Detail of Wiechert seismograph from 1902.

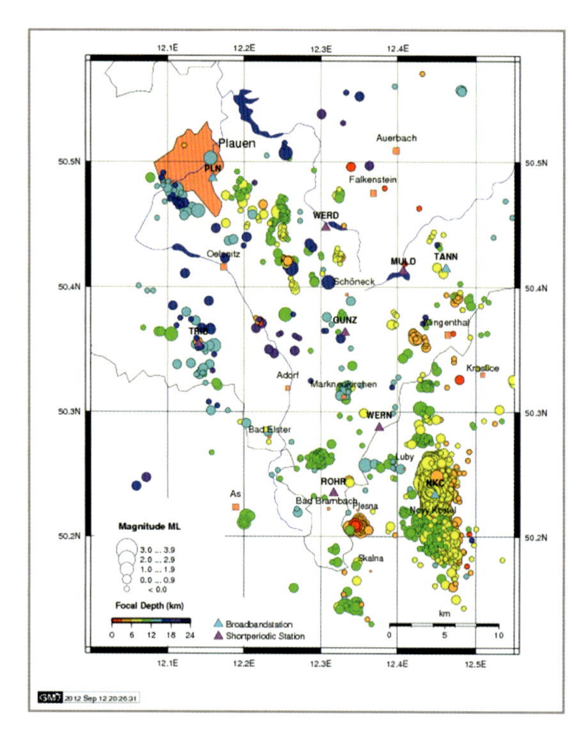

Figure 9.9: Seismicity in Saxony and adjacent areas as recorded by the West Saxonian Seismic Network.

meters from Jena, and other instruments recreated in the observatory's own precision mechanical workshop (Figure 9.8). The analogue data produced was recorded in the basement of the main building on photographic paper. Observations of global and local seismicity were analyzed manually and passed on to national and international data centres such as the International Seismological Center (ISC), from where they can be freely accessed for international research. Each year, the observatory analyses about 4500 global seismic events.

In 1993, the observatory was equipped with a STS-2 broadband sensor with digital data logging and online data transmission, bringing it up to the latest standards of modern seismological data acquisition, and since then

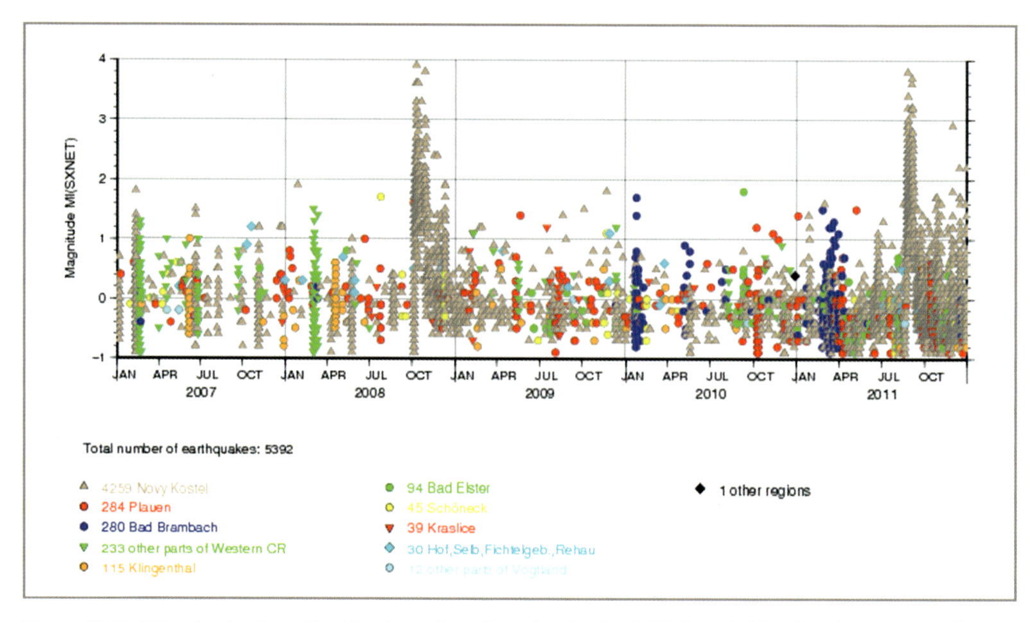

Figure 9.10: Magnitude-time distribution of earthquakes in the W-Bohemia/Vogtland swarm earthquake region during 2007–2011. The large swarms in 2008 and 2011 are clearly visible.

the observatory has been part of the German Regional Seismological Network (GRSN). The improved quality of the data with respect to dynamic and frequency range and the hitherto unknown possibilities of automatic data exchange opened up new perspectives for research and earthquake monitoring.

In 2000, additional seismic stations began to be set up in western Saxony. Today, the Institute of Geophysics and Geology runs a network of 12 stations and de facto fulfils the role of an earthquake alert service for the state of Saxony. About 600 small earthquakes are recorded in Central Germany each year (Fig. 9.9). However, during earthquake swarm activities in W-Bohemia/Vogtland (as in 2000, 2008 and 2011), this number rises to several tens of thousands (Fig. 9.10).

A new underground seismometer vault was constructed in 2006 (Fig. 9.11), and the recording equipment was moved in early in 2007. Thanks to the very stable conditions in the vault, the quality of the seismic measurements were further improved. The old seismometer hut ("Credner-Weickmann-Erdbebenwarte") with the historical instruments – some still operational today – is now open to the public, and can be visited by appointment. 400 visitors came in 2011, showing that the observatory also serves as an important tool for public outreach.

Figure 9.11: Construction of the new seismic vault in 2006.

The Zingst Maritime Observatory

10

Peter Hupfer and Hans-Jürgen Schönfeldt

Foundation

When Karl Schneider-Carius and Erich Bruns (Fig. 4.4; see Brosin 2001) met in spring 1956, their common interest was to develop a maritime branch at the Geophysical Institute. While Schneider-Carius, the newly appointed Director, had in mind that all of the three main fields of geophysics should be represented at his institute, Bruns was at the time seeking to establish an university-based department of physical oceanography. He had worked out a proposal, and it was quickly implemented. It was clear that a location at the coast was needed for this new division of teaching and research. A suitable object was found by H. v. Petersson (Fig. 10.1; see Hupfer 2003/04) at Müggenburger Weg 5 in the Baltic Sea resort of Zingst (Fig. 10.2), the lease of which ran out in 1956. The house had been built in 1937 as a residence for the commander of the air force training area that existed at the time and so it still belonged to the East German Ministry of Defence. Bruns had the appropriate connections and managed to have the responsibility for the property transferred to the University of Leipzig. In turn, Schneider-Carius took the necessary steps to obtain approval from the university to set up this new part of the institute. It should be emphasized that the administration of the university emphatically supported what was an unusual plan for Leipzig from the outset. A modest staff consisting of a lecturer (H. von Petersson), who was in charge locally, and two technical workers was approved. January 1, 1957 is considered to be the founding date for the Zingst Maritime Observatory of the Geophysical Institute. But Bruns withdrew from the project as early as 1958, when he saw the possibility of transforming the Hydro-Meteorological Institute of the Sea Hydrographic Service of the GDR (SHD), of which he was the head, into the Institute of Oceanography (Institut für Meereskunde) Warnemünde at the German Academy of Sciences.

Figure 10.1: Hans v. Petersson (1906–1992).

Figure 10.2: The Zingst Maritime Observatory. It stands about 200 m from the beach (top) View from the street (bottom) View from the garden.

Development

Since the foundation of the observatory was not the result of a planned investment, it started life as an empty house. By about mid-1957, furniture and the equipment necessary for teaching activities became available. Basic meteorological and oceanographic equipment was drawn from the holdings of the institute and of SHD. An open motor boat and a rowing boat were procured, as well as device winches and other equipment. In Leipzig, an assistant (Peter Hupfer) was appointed responsible for the maritime branch in August 1957.

In 1958, the house was renovated inside, a modern hydro-chemical laboratory was built and technical equipment was installed. Structural improvements continued (Figures 10.3, 10.4, 10.5), including the construction of a small

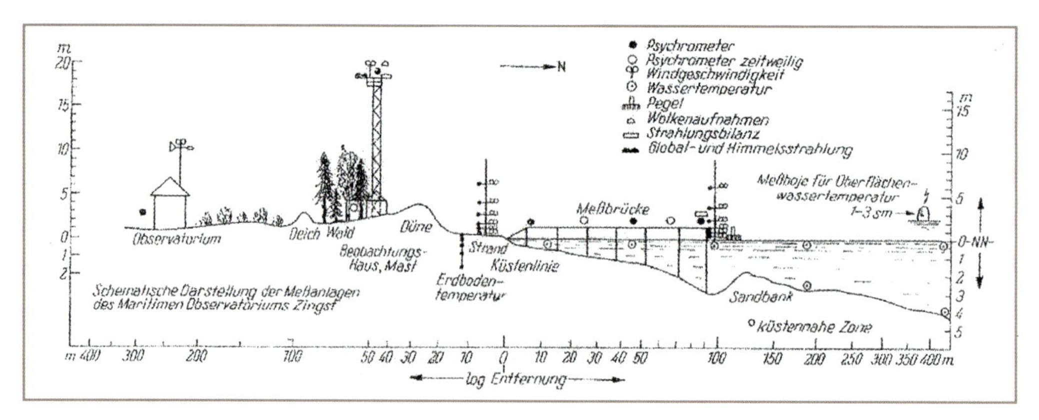

Figure 10.3: Schematic section of the measuring systems of the Zingst Maritime Observatory on both sides of the shore line (1960ies).

▲

Figure 10.5: Pier with gradient measuring masts (1970/80ies).

◀

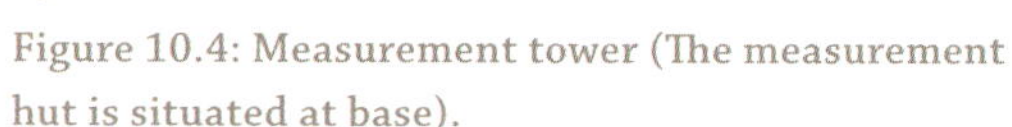

Figure 10.4: Measurement tower (The measurement hut is situated at base).

measurement hut in the dune area (1966), a 17-m tower (1965) for wind and radiation measurements, and an 80-m pier was built out into the sea for measuring purposes (1965). A solid little house was built in Zingst harbour at this time, serving primarily as a storage room for boat equipment and other devices. In 1970, the plot of land on which the observatory stood was expanded considerably (Hupfer und v. Petersson 1963).

In 1959, the first permanent assistant (Hans-Jürgen Brosin) and another technician were hired at the observatory. Adelheid Nitzschke and Hans-Ulrich Lass held the assistant's post later. In 1970, Henning Baudler was appointed as a further scientist. A milestone in the development was the acquisition of the research cutter "Atair" (Fig. 10.6), a rebuilt and modified 12-m fishing boat, in 1960. In 1973, "Atair" was replaced by the research vessel "Ikarus" (a former fire boat; Fig. 10.6).

Figure 10.6: (top) Research cutter "Altair" in Zingst (1969) (bottom) Research vessel "Ikarus" (1973).

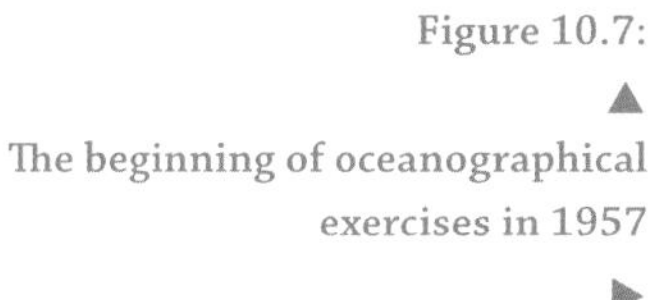

Figure 10.7:

▲

The beginning of oceanographical exercises in 1957

▶

H. v. Petersson and students during sea ice measurements in 1963.

The new field of activity made a leap in development in 1971. After the closure of the Geophysical Institute in Leipzig, several positions as well as the institute's mechanical and electronic workshop were transferred to the oceanography group remaining in Leipzig, which now bore the name *Working Group Oceanology and Maritime Meteorology*. Further development took place within the Department of Geophysics. The working group in Leipzig included Michael Börngen (1973), Thomas Foken (1974) and Armin Raabe (1975). The workshops were headed by the engineer Günther Neubert and the master precision mechanic Werner Bohmann. After the retirement of H. v. Petersson, the physicist Hans-Jürgen Schönfeldt began to work in the observatory in 1973, and in 1975 he was entrusted with the management of the observatory as its curator. After the appointment of Peter Hupfer to Berlin, Christian Hänsel became responsible for the working group as a whole.

The observatory saw itself from the outset as open to cooperation with and support for other groups. In consequence, there were productive cooperation arrangements with the Institute of Oceanography, the staff of the Institute for Physical Hydrography in Berlin and its successor, the Water Management Directorate Coast-Warnow-Peene, and with various other university institutes. Special assistance was provided to the Section Biology at Rostock University with the offer of all available resources until it opened its own Biological Station in Zingst in 1977.

Teaching and Promotion of young Scientists

Two- or three-week courses in oceanography and maritime meteorology were run by the observatory, the first of them even in 1957. Up to 14 participants could be accommodated in the house; they had usually attended an introductory lecture in oceanography before coming to Zingst. In addition to conveying knowledge in the form of lectures and seminars, teaching at the observatory put particular emphasis on the work done autonomously by the students. The courses in Zingst were popular with students and were generally an unforgettable experience (Fig. 10.7).

From the beginning, the observatory was open for all interested study programmes in the country to use at no charge. In the course of time, meteorologists, geophysicists, biologists, geographers and fishery scientists from all over East Germany completed courses in Zingst, and foreign student groups were also welcomed as guests.

Numerous *Diplom* theses were completed by meteorologists, physicists and teaching students at the Zingst Maritime Observatory, and a number of young scientists developed the foundations for their doctoral dissertations in Zingst.

P. Hupfer, also author of popular science books (Fig. 10.8), gave guest lectures at the universities of Berlin and Rostock, at the Technical University of Dresden and the Mining Academy in Freiberg, and was invited to give talks in Moscow and Gdańsk.

Figure 10.8: This popular book on the Baltic Sea was published from 1978–1984 in four editions.

Today, the Institute for Meteorology still organises two-week courses at the observatory (now called "Außenstelle Zingst" of the University of Leipzig) for students of meteorology.

Research

The first proposed line of research was the study of the bay south of Darß and Zingst, which was little explored at the time. This bay, or *Bodden*, is an almost enclosed body of water with a narrow connection to the Baltic Sea and is characteristic of the coast of Vorpommern. Experimental studies on the contact zone

Figure 10.9: Splashdown of the meteorological Coastal Radio Buoy KFB 2. Wind velocity, air temperature, humidity and sea surface temperature have been measured and transmitted to a land station (about 1975).

between land and sea was the other proposed direction of investigation, which later determined the observatory's research profile. Further areas of focus were on the long-term development of oceanographic and meteorological parameters around the Baltic Sea, and on selected issues of Baltic Sea Oceanography in Leipzig. After 1957, it became possible to participate in expeditions on Soviet research vessels (Hupfer, Brosin, Raabe).

In the research in the *Bodden*, a basic study of the water balance and hydrographic conditions was carried out in the years 1959–1965 by H.-J. Brosin. The results still today form the basis for multidisciplinary research into these water bodies. In addition, sporadic investigation focused on hydrographic fronts in these tideless estuaries, as well as on the spatial and temporal variability of salinity within them. The observatory took part in the SYNOPTA joint projects in the years 1972 and 1979. In the 1990s, a hydro-numerical model was applied successfully in relation to the storm surge of November 1995 (Hupfer, Schönfeldt, in press).

Investigations of the land/sea contact zone began in 1963 and continued until the early 1990s in addition to the *Bodden* research. Such studies, which were then a relatively new field, can be associated with microoceanography and micrometeorology. They ultimately aimed at gaining insight into the interactions between the sea and the atmosphere under the conditions of immediate coastal proximity. The investigation began with research on the heat budget and the temperature field in the near-shore zone (Hupfer). Intensive observations of the transition behaviour of near-surface meteorological properties on a spatial scale of 100 m followed (Nitzschke). It was demonstrated that by means of high resolution water temperature measurements in an appropriate array, estimates of the horizontal flow vector could be obtained (Baudler). H.-U. Lass studied the currents in the near-shore zone, which resulted in the development of an advection model within the framework of a research contract. For measurements further offshore, telemetric buoys (KFB; Fig. 10.9) were developed by G. Neubert and his co-workers in the 1970s. Wind direction and wind speed, air temperature and humidity as well as sea surface temperature were recorded and transmitted to a coastal station. A fall probe ("Fallsonde") has been developed, which is able to measure the molecular temperature boundary layer over water and other surfaces and thus provide measurements of the sensible heat flux (Foken).

Over a longer period, the main body of research activity was directed at investigating the characteristics of the wind field close to the ground near the shore. The internal boundary layer in the vertical distribution of wind, which caused by abrupt changes in surface properties, was measured by means of various experimental arrangements (Raabe, Schönfeldt and others).

Figure 10.10: Volume with the results of the International EKAM-Experiment (1973).

The development of wave probes began in the 1980s, and gave rise to a computerized data acquisition system. The data was used to verify wave model predictions for the coastal zone and to confirm theoretical models of edge wave characteristics (Schütt, Schönfeldt). The work was supported by the Federal Ministry for Education and Research BMBF and the Land Mecklenburg-Vorpommern after 1990. A wind atlas of the Bodden area was compiled (Hinneburg) within the framework of the *Climate change and Bodden area* (KLIBO) project, as along with a wave atlas and a sediment transport atlas of the outer Fischland, Darß, Zingst and Hiddensee coast (Riechmann, Stephan). Work was also done on aeolian sand transport (Schönfeldt).

One line of research of the working group, pursued until the late 1970s, was the long-term study of oceanographic and meteorological parameters in the Western Baltic Sea. Other research activity included analysis of strong salt water inflows into the Baltic Sea and the Sea's discharge climatology (Börngen, Hupfer).

International cooperation became more frequent and intensive. Some highlights were "Coastal Experiments" in the World Ocean programme of the Council for Mutual Economic Assistance (COMECON) with coordinated measurement of meteorological parameters, waves and currents, sediment movements, exchange of energy between sea and atmosphere, and others. The first experiment was carried out in Zingst in 1973 under the name EKAM'73 (influences of the coast on atmosphere and sea; Fig. 10.10) and included scientists from Poland, Russia, and the GDR. More coastal experiments were carried out in 1974 and 1976 on the Polish coast, and yet more in 1977 and 1979 on the Bulgarian coast. In addition to these big projects, cooperation with other working groups and individual scientists continued throughout the period.

A review of all the activities of the working group is given in Hupfer et al. (2005).

The University's multi-purpose Building

The Zingst Maritime Observatory enjoyed a further period of prosperity in teaching and research in the years immediately after 1990. The material as well as personnel capacities were expanded thanks to government support. The existing research boat could be replaced and modern equipment for inshore use could be acquired and put into use. While previous cooperative relations continued, new close collaborations with relevant institutions also developed; those with the Research Centre Geesthacht and the University of Hannover deserve particular mention.

When the Institute for Meteorology was founded in 1993, the working group was restructured and recent activities at the observatory came to an end. There are no plans to permanently staff the house, which is now called Außenstelle Zingst der Universität Leipzig (Zingst Site of the University of Leipzig) and is used for various purposes, including offering courses for students of meteorology.

As conceived by its originators and used in teaching and research, the Zingst Maritime Observatory was an institution unique in Germany and beyond. There can be no doubt that it would have maintained its position in the field of German maritime education and research if it could have been placed under the auspices of a suitable organisation.

References

d'Arrest, H. L. (1851): Bestimmung der Declination im magnetischen Observatorium zu Leipzig. Ber. Verh. Kgl. Sächs. Ges. Wiss, math.-phys. Classe, Vol. 2, pp. 100–105

Bergeron, T. and G. Swoboda (1924): Wellen und Wirbel an einer quasistationären Grenzfläche über Europa. Analyse der Wetterperiode 9–14. Oktober 1923. Veröff. Geophys. Inst. Univ Leipzig, 2. Ser., Vol. 3, pp. 63–172

Bjerknes, V. (1938): Leipzig – Bergen. Festvortrag zur 25-Jahrfeier des Geophysikalischen Instituts der Universität Leipzig. Z. f. Geophysik Vol. 14, Nr. 3/4, S. 49–62

Brosin, H.-J. (2001): Erich Bruns und das Institut für Meereskunde Warnemünde. Historisch-Meereskundliches Jahrbuch (Stralsund), Vol. 8, pp. 71–82

Bruns, H. (1878): Die Figur der Erde. Ein Beitrag zur europäischen Gradmessung. Berlin: Stankiewicz. 49 S. (Publication d. Kgl. Preuss. Geodät. Institutes; [8])

Hesse, W. (1949): Das Geophysikalische Institut der Universität Leipzig und das Geophysikalische Observatorium Collm. Veröff. Geophys. Inst. Univ Leipzig, 2. Ser., Vol. 15, pp. 7–13. Leipzig

Hesse, W. [Ed.] (1961): Handbuch der Aerologie. Akad. Verlagsgesellschaft Geest & Portig, Leipzig, 897 S.

Hesselberg, T. and A. Friedmann (1914): Die Größenordnung der meteorologischen Elemente und ihrer räumlichen und zeitlichen Ableitungen. Veröff. Geophys. Inst. Univ Leipzig, 2. Ser., Vol. 1, pp. 147–173

Hupfer, P. (2003/04): Seemann und Forscher – eine Erinnerung an Hans von Petersson anlässlich der 100. Wiederkehr seines Geburtstages. Historisch-Meereskundliches Jahrbuch (Stralsund), Vol. 10, pp. 29–38

Hupfer, P. (2012): Die Institute für die Meteorologenausbildung in der DDR und ihre Auflösung 1971 und 1996. Manuskript, 15 p.

Hupfer, P. und H. v. Petersson (1963): Das Maritime Observatorium Zingst des Geophysikalischen Instituts der Karl-Marx-Universität Leipzig. Veröff. Geophys. Inst. Univ Leipzig, 2. Ser., Vol. 18, pp. 35–56

Hupfer, P., H.-J. Schönfeldt und A. Raabe (2005): Das Maritime Observatorium Zingst der Universität Leipzig 1957–1994. Historisch-Meereskundliches Jahrbuch (Stralsund), Vol. 11, pp. 39–72

Hupfer, P. and H.-J. Schönfeldt: Boddenforschung durch das Maritime Observatorium Zingst der Universität Leipzig. Historisch-Meereskundliches Jahrbuch (Stralsund), in press.

Koch, H. [Ed.] (1969): Industriemeteorologie. Eine Sammlung von Beiträgen zu Fragen der angewandten Meteorologie in der Industrie. Eigenverlag der Karl-Marx-Universität Leipzig, Leipzig, 419 S.

Kortüm, Fr. (1963): Das Geophysikalische Institut der Karl-Marx-Universität in der Zeit von 1949 bis 1962. Veröff. Geophys. Inst. Univ Leipzig, 2. Ser., Vol. 18, pp. 7–16

Kortüm, Fr. (1966): Beiträge zur Klimatologie des Wärmehaushaltes der Erdoberfläche im norddeutschen Flachland. Habil.-Schrift, Math.-Naturwiss. Fakultät der Humboldt-Universität zu Berlin

Lammert, L. (1920): Der mittlere Zustand der Atmosphäre bei Südföhn. Veröff. Geophys. Inst. Univ Leipzig, 2. Ser., Vol. 2, H. 7, S. 261–323

Lammert, L. (1932): Das Geophysikalische Institut der Universität Leipzig von seiner Gründung im Jahre 1913 an bis zum Jahre 1932. In: Herrn Prof. Dr. L. Weickmann zum 50. Geburtstag. (Unpublished, without pagination)

Lammert, L. and M. Dietsch (1921): Normalwind und Reibungskraft in 1000 m Höhe. Beitr. z. Physik d, freien Atmosphäre, Vol. IX, pp. 67–77

Nitzschke, P. (1966): Die Temperaturverhältnisse in der freien Atmosphäre über dem Südpolargebiet. Veröff. Geophys. Inst. Univ Leipzig, 2. Ser., Vol. 19, pp. I–X, 1–63

Philipps, H. (1962): Prof. Ludwig Weickmann in memoriam. Z. f. Meterol. Vol. 16, pp. 121–126, with list of publications

Schneider-Carius, K. (1961): Das Klima, seine Definition und Darstellung: Zwei Grundsatzfragen in der Klimatologie. Veröff. Geophys. Inst. Univ Leipzig, 2. Ser., Vol. 17, pp.149–222

Sverdrup, H. U. (1951): Vilhelm Bjerknes in Memoriam. Tellus, Vol 3, Nr. 4 (Nov 1951), pp. 217–221

Weickmann, L. (1938): Bericht zur 25-Jahr-Feier des Geophysikalischen Instituts der Universität Leipzig. Das Geophysikalische Observatorium. Vorwort. Veröff. Geophys. Inst. Univ Leipzig, 2. Ser., Vol. 10, pp. 7–14. Leipzig

Archive Material:

Universitätsarchiv Leipzig, PA 1159

Chronicle

1409	2 December: Foundation of the University of Leipzig
1913	1 January: foundation of the Geophysical Institute. Vilhelm Bjerknes is first Professor of Geophysics and Director of the Institute located at Nürnberger Straße 57
1917	Summer: Geophysical Institute moves to Talstraße 38
1917	After Bjerknes' return to Norway, Wenger becomes Director of the Institute
1922	19 September: foundation of the German Seismological Society (after 1924 German Geophysical Society) in Talstraße 38
1923	Ludwig Weickmann becomes Professor of Geophysics and Director of the Geophysical Institute
1932	6 October: the Collm Geophysical Observatory inaugurated
1943	4 December: the building at Talstraße 38 is destroyed; the Institute moves temporarily into Talstraße 35
1945	23 June: before Leipzig is incorporated into the Soviet Occupation Zone, Ludwig Weickmann and various other scientists place the University of Leipzig under American control
1950	Max Robitzsch becomes Professor of Geophysics and Director of the Geophysical Institute and the Collm Geophysical Observatory
1951	Geophysical Institute moves to Schillerstraße 6
1953	5 May: The University of Leipzig is renamed "Karl-Marx-Universität" (until 1991)
1955	Karl Schneider-Carius becomes Professor of Meteorology and Director of the Geophysical Institute
1957	1 January: Zingst Maritime Observatory founded
1957	1 September: Robert Lauterbach becomes Professor of Applied Geophysics
1958	1 March: Institute for Geophysical Investigation is established in Talstraße 35
1965	Institute for Geophysical Investigation expands to incorporate the Geological-Palaeontological Institute, forming the Institute for Geophysical Investigation and Geology
1969	1 February: the Institute for Geophysical Investigation and Geology is transferred to the Department of Geophysics (Talstraße 35) within the Section of Physics
1971	Closure of the Geophysical Institute. The two observatories and the most Institute staff are incorporated in the Department of Geophysics of the Physics Section. The students and the rest of the staff transfer to Humboldt University Berlin
1990	Unofficial establishment of the Institute of Geophysics, Geology and Meteorology
1993	2 December: Foundation of the Institute of Geophysics and Geology under Franz Jacobs (Talstraße 35) and the Institute for Meteorology under Gerd Tetzlaff (LIM, Stephanstraße 3). The new institutes are incorporated into the Faculty of Physics and Earth Sciences
1994	Zingst Maritime Observatory is closed
2004	Werner Ehrmann becomes Director of the Institute of Geophysics and Geology
2008	Manfred Wendisch becomes Director of the Institute for Meteorology

Personnel Register

Further Information

The present book describes some specific aspects of Leipzig's contributions to geophysics and meteorology over the last 100 years. More detailed information on the topic can be found at http://www.geo.uni-leipzig.de/. The web page includes a collection of other related publications, such as graduation lists and tables of contents of doctoral and habilitation theses.

Figure Sources

Archiv Börngen: Title (stamp), 1.7, 2.3, 4.8, 5.3, 6.2, 6.6, 6.8, 10.8, 10.10; Archiv Hupfer: 3.1, 3.5, 4.1–4.3, 4.5, 4.7, 10.2, 10.5, 10.6; Archiv Jacobs: 5.10, 7.2; Archiv v. Petersson: 10.7; Archiv Schellenberg: 3.4; Archiv Tetzlaff: 2.18; Archiv Weickmann: 2.5, 2.9, 2.13, 2.16, 2.17; d'Arrest 1851, p. 104: 1.1; Beitr. z. Physik d. freien Atmosph., Vol. 19 (1932): 1.6; Ber. dt. Ges. geol. Wiss., Reihe A, Vol. 12, 1967: 5.1; Bergakademie Freiberg: 2.12; Bergeron, Swoboda (1924), p. 72: 2.4; DGG-Mitteilungen, Sb 1/2001, p. 57: 5.9; Hupfer 2003/04, p. 35: 10.1; Hupfer et at al. 2005, p. 16: 10.3; Ibérica, Vol. 28 (1927), No. 694, p. 169: 2.7; Institut für Geophysik und Geologie: Title (Collm), Title (Merapi), 1.3, 1.8, 1.9, 1.11, 1.12, 2.6, 3.2, 5.2, 5.4, 5.6–5.8, 5.11, 5.12, 6.3, 6.4, 6.7, 6.10–6.13, 7.1, 7.3–7.11, 9.1–9.3, 9.8–9.11, 10.4; Institut für Meteorologie: Title (aerosol effect), 1.10, 2.8, 6.9, 8.1, 8.3, 8.5, 9.6, 9.7, 10.9; Lauterbach, Wiss. Z. d. Karl-Marx-Univ. Leipzig, Vol. 5 (1955/56), p. 518: 5.5; Leibniz-Institut für Länderkunde, Leipzig: 2.1; 50 Jahre Geophysik in Leipzig (2001), p. 13: 6.1; Nachruf auf Otto Wiener, Ber. math.-phys. Kl. Sächs. Ak. Wiss., Vol. 79 (1927), p. 107: 1.5; Nitzschke 1966, Table I: 4.6; Nachruf Prof. Dr. Walter Hesse, Z. Meteorol., Vol. 32 (1982), p. 197: 3.3; Pospichal: 8.2; Universitätsarchiv Leipzig: 2.2, 2.10, 2.15; Universitätsbibliothek Leipzig: 1.4, 2.11, 2.14, 3.6, 4.4, 6.5, 8.4, 9.5; Veröff. Geophys. Instituts, 2. Serie, Vol. 10, p. 22: 9.4; Wahnschaffe: Hermann Credner. Z. dt. geol. Ges., 65: Monatsber., 8-10: 1.2

List of Staff

(No requirement of completeness)

Geophysical Institute 1913–1971, Institute for Geophysical Investigation (and Geology) 1958–1969, Section of Physics – Department of Geophysics 1969–1993, Institute of Geophysics and Geology 1993–2012, Institute for Meteorology 1993–2012.

Jens **Adam** Alfred **Adlung** M. Fouzou **Al-Habash** Barbara **Angelmi** Bernadette **Antkowiak** Erika **Apitz-Lück** Ernst Moritz **Arndt** Klaus **Arnold** Lars **Aschmann** Clemens **Auerbach** Henry **Auraß** Frank **Bach** Kathi **Balogh** Georg **Banz** Manuela **Barth** Henning **Baudler** Klaus **Bauer** Stefan **Bauer** W. **Becker** Björn R. **Beckmann** H. **Behrend** Michael **Bender** Tor **Bergeron** Kathrin **Bergmann** Maritta **Berndt** F. **Bernhardt** Karl-Heinz **Bernhardt** Andreas **Berthold** Andreas **Beser** Hildegard **Beyer** Fritz Holm **Bielich** Karin **Bienert** Eike **Bierwirth** Jacob **Bjerknes** Vilhelm **Bjerknes** Melanie **Bock** Áron **Böcker** J. **Boddin** C. **Boerner** Werner **Bohmann** Madelaine **Böhme** Otto **Bohne** Antje **Böhnke** R. **Böker** Andre **Bornemann** Frieda **Börner** Michael **Börngen** Felix **Böthig** Thomas **Brachert** Johannes **Bracht** Theodor **Brandes** Ute **Brauer** Hans-Jürgen **Brosin** Marlen **Brückner** Ingolf **Brunner** Erich **Bruns** Petra **Buchholz** Holger **Bürkholz** Edda **Clauß** Jürgen **Clauß** Sven **Damm** Erik **Danckwardt** M. **Deckert** H. **Deinhardt** Rene **Devantier** Angelika **Dick** Kurt **Diesing** Marie **Dietsch** Gerhard **Dietze** Götz **Dittrich** Hildegard **Domschke** Mrs. **Donow** Sylvia **Dorn** Willy **Duckheim** Andre **Ehrlich** Werner **Ehrmann** Heidi **Eichhorn** Lothar **Eißmann** Manfred **Engelhorn** Erich **Etienne** Mohammad **Fallahi** Oliver **Fanenbruck** Gerhard **Fanselau** Robin **Faulwetter** Wolfgang **Feck-Yao** Fanni **Finger** Gabi **Fischer** Christina **Flechsig** Hortencia **Flores** Thomas **Foken** Veronica **Frank** Daniel **Franke** Ursula **Franz** Klaus **Freitag** Clemens **Fricke** Diethard **Fricke** Sven **Friedel** Christian **Friedrich** Sieglinde **Friedrich** Jens-Uwe **Friemann** Christina **Fröhlich** Michael **Fuchs** Sigward **Funke** Peter **Gäbler** Lisa **Gaitzsch** Martina **Gassenmeier** Leo **Gburek** Margot **Geißler** Walter **Gläßer** Birgitt **Göbel** Hein **Graber** Frank **Grasemann** Steffen **Grässl** Olaf **Graupner** Helmut **Greulich** K. **Grießbach** Annelore **Grimmer** Dieter **Groß** Johannes **Großer** Sabine **Grummt** Gottfried **Grünthal** Christoph **Grützner** Daniel **Günther** Bernd **Haacke** Herbert **Haase** Rudolf **Haase** Yvonne **Hamann** Christa **Hanke** Christian **Hänsel** Uwe **Harlander** Silke **Hartmann** Wolfgang **Haupt** Bernhard **Haurwitz** Olaf **Hein** Jens **Heinicke** Harald **Heinrich** Robert **Heinse** Eduard **Heise** Alfred **Helbig** Sabine **Helbig** Olaf **Hellmuth** Matthias **Henniger** H. **Henning** Andreas **Herber** Wolf-Dieter **Hermichen** Martin **Herrmann** Walter **Hesse** Harald **Hessel** Theodor **Hesselberg** Marc **Heydecke** Lothar **Hiersemann** Rolf **Hillebrand** Detlef **Hinneburg** Hans **Hinzpeter** Kerstin **Hirsch** Silke **Hock** Stefan **Hodam** Kerstin **Hoffmann** Peter **Hoffmann** Roland **Hohberg** Hannelore **Höhne** J. **Holtsmark** Elke **Hoppe** Ursula **Hotzelmann** Wolfgang von **Hoyningen-Huene** Katja **Hungershöfer** Peter **Hupfer** Eckart **Hurtig** Mr. **Israel** Christoph **Jacobi** Franz **Jacobs** Angela **Jaeschke** Evelyn **Jäkel** Antje **Jordan-Heydecke** Tamara **Judersleben** Christof **Junge** Frank **Junge** Olaf **Juschus** Anita **Just** Günther **Just** Falk **Kaiser** Pham **Khoan** Torsten **Kildal** Ray **Klimmek** Ursula **Klingner** Martin **Klingspohn** Norbert **Klitzsch** Gottfried **Kluge** Rudolf **Knieß** Anke **Kniffka** Siegfried **Knoth** Hans **Koch** Mathias **Koch** Elsa **König** Uwe **Koppelt** Michael **Korn** Friedrich **Kortüm** Christina **Krause** Yvonne **Krause** E. **Kregel** Stefan **Krüger** Ute **Krüger** Siegfried **Kuhn** Tanja **Kuhnt** Johannes **Kulenkampff** Thorolf **Küpper** Dierk **Kürschner** Dana

Laaß-Kießling Luise **Lammert** Jan-Michael **Lange** Martin **Lange** Hans-Ulrich **Lass** Robert **Lauterbach** Ronny **Leder** Eva **Lehmann** Heinz **Lettau** Dirk **Leuschner** Martha **Lindner** Thomas **Litt** Renqiang **Liu** Rainer **Löbe** Wilhelm **Löhning** Martin **Lösche** Ernst **Maasch** Rosemarie **Maaß** Andreas **Macke** Albert **Mäde** Peter **Mayer** Magdalena **Meissner** Martin **Melles** Karsten **Menschner** Hans **Merkel** Georg **Merkler** Werner **Metz** Britta **Mey** Cornelius **Meyer** Helmut **Meyer** Ursula **Mickan** Paul **Mildner** Fritz **Model** Emile **Moise** Nicole **Mölders** F. **Möller** Klaus **Morsch** Seyede-Sima **Mousavi** Manfred **Mudelsee** Manuela **Mühl** Annelore **Müller** Arnold **Müller** Christine **Müller** J. **Müller** Ulrich **Müller** K. **Munzig** Werner **Nagel** Günther **Neubert** Marco **Neumann** Walter **Neumann** Hans **Neumeister** Walter **Neustadt** Bernd **Nikutowski** Hildegard **Nitsche-Kohlsche** Adelheid **Nitzschke** Peter **Nitzschke** Gerhard **Noßke** Gerd **Olszak** Peter **Opitz** Sabrina **Ortlepp** Guido **Öttinger** Sebastian **Otto** Hendrik **Paasche** Werner **Pagels** Klaus **Pardeyke** Ingrid **Peifer** William **Peine** Rudolf **Penndorf** Gerald **Peschel** Hans **von Petersson** Günter **Petzold** H. **Petzold** Hannelore **Pfuhl** Helmut **Piazena** Josepha **Pietrzak** Jan **Piotrowski** Hans-Jürgen **Pitzschel** Matthias **Plehn** Kerstin **Polozek** Eraldo **Pomponi** Rudolf **Ponndorf** Hans-Helmut **Poppitz** Bernhard **Pospichal** Peter **Posse** Iris **Proeck** Jens **Przybilla** Johannes **Quaas** Armin **Raabe** Kai **Radtke** Horst **Rast** Jutta **Rehnert** Klaus **Reicherter** D. **Reinhardt** Silvia **Reiniger** Claudia **Reißmann-Schütze** Charlotte **Richter** Frank **Riechmann** Ulrich **Rindelhardt** Josef **Rink** Hans **Rische** Max **Robitzsch** Javier **Rojas** Dagmar **Rosenow** Dirk **Rößler** Tillmann **Roth** Aleksandr **Rozenberg** Karsten **Rücker** Albrecht **Saalbach** A. V. R. **Sastry** Michael **Schäfer** Horst **Schellenberg** Gerwalt **Schied** Sigurd **Schienbein** Peter **Schikowsky** Uwe **Schlink** Andreas **Schmidt** G. **Schmidt** Steffi **Schmidt** Gerhard **Schmiedl** Rita **Schminder** Rudolf **Schminder** Karl **Schneider-Carius** Volkmar **Schnieber** Dieter **Schöne** Eberhard **von Schönermark** Maria **von Schönermark** Hans-Jürgen **Schönfeldt** Annemarieke **Schoonderwaldt** Peter **Schreck** Kurt **Schubert** Klaus **Schütt** A. **Schütz** Christine **Schwabe** Gerhard **Schwarz** Dorothea **Schwarze** Martin **Seidel** G. **Seifert** Manfred **Seifert** Gerda **Seifert-Weber** Ch. **Seltz** Christoph **Sens-Schönfelder** Ulrich **Serfling** Dirk **Siekiera** Gert **Siewert** Martin **Simmel** Heinz **Sinz** Jörg **Skamletz** Mrs. **Sohr** Halvor **Solberg** Reiner **Sommerweiß** Dietrich **Sonntag** Heide **Spiegelberg-Storz** Renate **Spindler** Karin **Staake** Anke **Steinbach** Mirko **Stephan** Ch. **Steyer** Gunter **Stober** Oliver **Stock** Karina **Stockmann** Claudia **Stolle** Wolfgang **Storz** Anne-Babett **Stoss-Ernst** H. **Stoye** Dietrich **Stranz** Gerhard **Streidt** Harald Ulrik **Sverdrup** Ina **Tegen** Gerd **Tetzlaff** Gundolf **Thiele** E. **Thiem** Helmuth **Thierbach** Ralf **Tietz** Bernd **Tittel** Thomas **Trautmann** E. **Truöl** Antje **Tuch** Burkhard **Ullrich** Claudia **Unglaub** Holger **Utke** Adam **Vecsei** Gerhard **Vogler** Horst **Voigt** Rene **Voigt** Bernd **Wagner** Hans **Wallenhauer** Andreas **Walter** Jörg **Walter** Thorsten **Wanger** Wolfgang **Warmbt** Jochen **Wauer** Friedhold **Weber** Ulrich **Wegler** Gisela **Weichsel** Ludwig F. **Weickmann** Helga **Weintritt** Konrad **Weißbrodt** Frank **Weiße** Georg **Weller** Manfred **Wendisch** Siegfried **Wendt** Robert **Wenger** Volker **Wennrich** Th. **Wereide** Frank **Werner** Melitta **Werner** Erwin **Wiechert** Madlen **Wild** Wenke **Wilhelms** Michael **Wilsdorf** Elfriede **Winkler-Lange** Holger **Wirth** Thomas **Wolf** W. **Wünsche** Silke **Zander** Waltraud **Zech-Foken** Astrid **Ziemann** Janek **Zimmer** Henning **Zöllner** Sebastian **Zschunke**